Biomechanics

Edited by Hadi Mohammadi

Published in London, United Kingdom

IntechOpen

Supporting open minds since 2005

Biomechanics
http://dx.doi.org/10.5772/intechopen.73352
Edited by Hadi Mohammadi

Part of IntechOpen Book Series: Biomedical Engineering, Volume 4
Book Series Editor: Robert Koprowski

Contributors
Haisheng Yang, Sheik Abdullah A, Selvakumar S, Parkavi R, Suganya S, Hadi Mohammadi, Emily Earl, Victoria Spartacus, Irina Bubnova

First published in London, United Kingdom, 2019 by IntechOpen
IntechOpen is the global imprint of INTECHOPEN LIMITED, registered in England and Wales, registration number: 11086078, The Shard, 25th floor, 32 London Bridge Street
London, SE19SG - United Kingdom
Printed in Croatia

British Library Cataloguing-in-Publication Data
A catalogue record for this book is available from the British Library

Additional hard copies can be obtained from orders@intechopen.com

Biomechanics
Edited by Hadi Mohammadi
p. cm.
Print ISBN 978-1-78985-121-2
Online ISBN 978-1-78985-122-9
ISSN 2631-5343

For EU product safety concerns:
IN TECH d.o.o., Prolaz Marije Krucifikse Kozulić 3, 51000 Rijeka, Croatia, info@intechopen.com or visit our website at intechopen.com.

IntechOpen Book Series
Biomedical Engineering
Volume 4

Dr. Hadi Mohammadi received his BSc and MSc degrees in Mechanical Engineering from Sharif University of Technology, Tehran, Iran, and his PhD degree in Biomedical Engineering (Biomaterials) from the University of Western Ontario. He has been a postdoctoral trainee for almost four years at the University of Calgary and Harvard Medical School. Regarding his recent work in the area of cardiovascular mechanics and technology, he achieved a global first in applying the soft robotic concept to the construction of prosthetic heart valves. He designed the "Okanagan valve," the next generation of bileaflet mechanical heart valves. Also, as featured in CBC 2016, he created synthetic heart valves and grafts that could transform training for major cardiovascular surgeries. His team members are industry innovators having created the technology to produce lifelike synthetic platforms that can be used for the simulation of almost all cardiovascular reconstructive surgeries.

Editor of Volume 4:
Hadi Mohammadi
University of British Colombia, Kelowna, Canada

Robert Koprowski, MD (1997), PhD (2003), Habilitation (2015), is a lecturer at the Department of Biomedical Computer Systems, Institute of Computer Science, University of Silesia, Poland. For 20 years, he has been dealing with analysis and processing of biomedical images with a particular emphasis on the full automation of measurement for a large inter-individual variability of patients. He is the author of dozens of papers with an impact factor (IF) and more than a hundred of other papers as well as the author or coauthor of six books. Additionally, he is the author of several national and international patents in the field of biomedical devices and imaging. Since 2011, he has been a reviewer of grants and projects (including EU projects) in the field of biomedical engineering.

Book Series Editor:
Robert Koprowski
University of Silesia, Poland

Scope of the Series

Biomedical engineering is one of the fastest growing interdisciplinary branches of science and industry. The combination of electronics and computer science with biology and medicine has resulted in improved patient diagnosis, reduced rehabilitation time and better quality of life. Nowadays, all medical imaging devices, medical instruments or new laboratory techniques are the result of the cooperation of specialists in various fields. The series of biomedical engineering books covers such areas of knowledge as chemistry, physics, electronics, medicine and biology. This series is intended for doctors, engineers and scientists involved in biomedical engineering or those wanting to start working in this field.

Contents

Preface

The purpose of this book is to educate readers who have a background in engineering and science and are familiar with topics such as statics, dynamics, mechanics of materials, thermodynamics, fluid mechanics, etc. There is absolutely no need for readers to have a background in anatomy, physiology, or even biology. Every chapter begins with a brief assessment of the related biological background followed by identification and explanation of the necessary features of the associated biomechanics problems.

Multiple topics such as biomechanics of human blood, orthopedics, and risk analysis have been discussed and analyzed in a variety of applications. Biological processes are amazing in their complexity and optimization. Blood, being no exception, is extremely evolved and adapted to the different scenarios necessary to maintain life. Consisting of plasma, white blood cells, platelets, and red blood cells, it is able to transport vital molecules around the body, including oxygen and clots in the case of injury. Since red blood cells make up approximately half the volume of blood, blood flow mechanics are largely related to the properties of red blood cells defined under a soft solid. Large deformability is essential in the life cycle and function of red blood cells as capillaries are extremely small. Blood clotting is a very important function of blood. All these concepts are admirably discussed in one chapter and readers with different backgrounds will follow and learn in an effective manner. The same concept applies to other chapters.

We all know that bone is a hard biological tissue. It is made up of cells located in the bone matrix, which is made up of mainly collagen fibers, amorphous ground constituent, and a mineral phase. The key components of bone mineral are known to be calcium carbonate and calcium phosphate. The mineral constituents contain primarily hydroxyapatite crystals and amorphous calcium phosphate. It is very well stablished that bone is the main reservoir for calcium in the human body. It also protects important organs such as heart and liver, provides mechanical stability, and is responsible for locomotion and movement of the human body. Bone structure is known to be hierarchical and is intended to be effectively optimal. Bone implants are considered as medical devices meant to substitute or provide fixation, or to substitute articulating surfaces of a joint. In other words, bone implants are designed to either support or substitute broken or damaged joints and bones. Bone implants are primarily made up of titanium alloys and stainless steel for their strength and are often co-used with a particular polymeric material as artificial cartilage. The main job of these implants is to decrease the stress in the implants and at the articulating surfaces. There are implants that are cemented into place and some that are press-fit, in a way the bone tissue in the vicinity of the implant can grow into the implant, hence, providing more strength and stability. The main aspect that guides bone healing is known as the interfragmentary movement, which regulates tissue strain and therefore the cellular response in the healing zone. The chapters in this area discuss comprehensively all these concepts in plain language.

Finally, I would like to thank my exceptional student, Emily Earl, who has been helping me throughout intellectually and technically in editing this book.

Hadi Mohammadi, PhD, PEng
Assistant professor of mechanical and biomedical engineering and surgery,
The Heart Valve Performance Laboratory,
School of Engineering,
Faculty of Applied Science,
University of British Colombia,
Kelowna, Canada

Section 1

Introduction

Chapter 1

Introductory Chapter: Biomechanics

Hadi Mohammadi

1. Introduction

Biomechanics or biomechanical engineering is the application of mechanical engineering and its concepts and principles in biological systems, living tissues/organs, and medical devices. Mechanical properties of tissues can be characterized as anisotropy, hyperelasticity, viscoelasticity, viscoplasticity, preconditioning performance, and the existence of residual stresses [1]. Most experiments on the mechanical characterization of tissues are based on the laboratory work. Often, samples are removed from cadavers or animals and are cut in order to be tested which are either fresh or after storage. Testing machines are often based on electromechanical or hydraulic systems which are often performed in living animals or patients. Mathematical interpretation of data is often considered an important part of tissue mechanics such as 3D modeling of stress-strain behavior. Modeling may be either phenomenological, which is to some extent seeking to define behavior using model systems that do not reference structure, or plainly based on information of tissue construction. Phenomenological models are often based on linear or quasi-linear viscoelastic theory. Constitutive equations, mostly those based on improvement of strain energy density functions, are often as a means to explaining tissue behavior under random loading.

Tissue mechanics or tissue biomechanics is the field of endeavor that seeks to understand and describe the correlations between structure, composition, and mechanical functionality in a variety of tissues such as connective tissues, cardiovascular tissue, epithelial tissues, etc., in the human body [2]. Much of the research work has been done in the connective tissues of the body, such as the bone, tendons, cartilage, arteries, and skin, where mechanical demands are greatest; however, all tissues have mechanical features of interest. An interesting area of research could be made around the (1) use of structural anatomy as a means to understanding natural design, (2) mechanical engineering analysis of structures based on continuum mechanics, and (3) materials science study of detailed links between structure and function. As all natural tissues are composite materials, understanding their mechanical function requires study of the mechanical properties and architectural arrangement of the individual structural components, particularly strong, stiff collagen fibers; the physiological rubber elastin; hydroxyapatite mineral; and proteoglycan sol/gels. The mechanical features of tissues include marked anisotropy, nonlinear stress-strain relations, viscoelasticity, preconditioning behavior, and the presence of residual stresses. Most studies of the mechanical behavior of tissues have been carried out in the laboratory, with samples removed from cadavers or animals, cut or machined to shape, and tested either fresh or after storage. Commercial testing machines based on electromechanical or hydraulic systems are

IntechOpen

widely used, as are custom-built apparatus. Testing is also carried out in living animals or patients. In either case, determination of sample geometry or deformations is difficult. Mathematical description of data is an important part of tissue mechanics, as is modeling of 3D stress-strain behavior. Soft tissues require the use of large deformation (finite) elasticity equations. Modeling may either be phenomenological (seeking to describe behavior using model systems that do not reference structure) or may be explicitly based on knowledge of tissue architecture. Phenomenological models have often been based on linear or quasi-linear viscoelastic theory derived from previous work on polymer materials. Constitutive equations, particularly those based on the development of strain energy density functions, have been widely used as a means to describing tissue behavior under arbitrary loading. In a complementary development, finite element analysis has found wide use for analysis of complex structures. I believe tissue mechanics is a tool both for (1) the study of natural structures in health and disease and (2) technological application, which, in recent years, has included the design of surgical replacements and surgical technique and evaluation and design of tissue engineered replacements and analysis/prevention of injuries. Native and functioning tissues display an exciting and dynamic response to mechanical loading. In fact, living tissues are capable of adapting to the micromechanical environment surrounding them which may be characterized by changes in their composition and structure due to their external conditions. Following this concept, tissue differentiation and profanation are good examples in which they are highly dependent and significantly affected by on their micromechanical environment. A quantifiable understanding of these concepts, by the means of experimentation and/or computational modeling, is of fundamental significance for many applications in the area of biomechanics or biomechanical engineering such as tissue and cell damage due to continued loading (e.g., decubitus), the usage of suitable biomaterials in the design of prostheses (e.g., intervertebral disc or in small size blood vessel, i.e., coronary arteries), and heart valve tissue engineering [1].

As of the future research directions, developing computational multiscale models to study the micromechanical loadings on discrete cells in an engineered tissue such as cartilage tissue or heart valve leaflet tissue is of particular importance. This is because these models will help understand the process by which normal living cells detect and realize mechanical loadings such as shear forces or tensile strains and convert them into a multiple of chemical events in tissues at multiscale levels that trigger and affect the cellular function in health and disease [3, 4]. Living cells vigorously detect, realize, and process micromechanical signals surrounding them, and all their motility, differentiation, and growth are significantly affected by them. It is of particular importance to understand the micromechanical language between the extracellular matrix (ECM) and cells embedded in it and the fundamental mechanisms of this mechanical communication simply because an abnormality in the mechanical properties of the ECM may lead to numerous diseases, such as fibrosis and cancer. Furthermore, ECM mechanics plays a major role when stem cell differentiation is programmed for organ-on-chip applications. Mechanotransduction is a developing multidisciplinary area that embraces cell and developmental biology, biomaterials, biochemistry, biomedical engineering, and medical biophysics [5, 6].
The following topics are a few applications of biomechanics in order to further understand the mechanobiology associated with tissues in health and disease.

2. Spine biomechanics

To understand the motion of the spine and how it supports movement, we must first understand each component and the overall structure of the spine. The spine serves as a structural column providing the human body physical durability and protection. There are 32–34 total bones in the spine, divided into 5 regions: 7 vertebrae in the cervical region, 12 in the thoracic region, 5 in the lumbar region, 5 in the sacral region, and 3–5 in the coccygeal region. In between these spinal bones, there are 23 discs. The bones and discs, in combination with muscles and ligaments, allow the lumbar, thoracic, and cervical spine, different degrees of mobility. This is measured through rotational, side to side, and front to back bending. As mobile humans, it is important to realize the impact of certain activities on the spine, in order to avoid injuries and spinal conditions. Looking at the spine as a purely mechanical system, solutions can be identified using engineering principles. By making use of medical professionals' expertise in surgical procedures and the human body and engineers expertise in mechanical systems and hardware, spinal disorders can be treated. The motion of the spine is very complex, and in order to fully understand the biomechanics of this system, the components, motion, force transmission, conditions, and treatment of the spine must be examined. Understanding the motion of the spine and how it supports movement eases the understanding of each component and the overall structure of the spine. The spine serves as a structural column providing the human body physical durability and protection. The bones and discs, in combination with muscles and ligaments, allow the lumbar, thoracic, and cervical spine to move with different degrees of mobility. This is measured through rotational and lateral bending and flexion/extension. As discussed the motion of the spine is complex. By fully understanding the biomechanics of the spine as a mechanical system, the components, motion, force transmission, and conditions of the spine can be harnessed to develop new technologies in vibration reduction, sports performance, and new methods and practices for spinal surgery.

3. Articular cartilage biomechanics

Cartilage is an essential component of the human body. It plays multiple roles throughout the body, and without it our bodies would not have the ability to respond to the demanding nature of our everyday lives. Cartilage is the basis of skeletal growth; it transmits and cushions demanding loads throughout brittle bone structures. Also, it gives elasticity and shape to surrounding tissues, all this while being an avascular tissue. The complexity of cartilage's biomechanical behavior makes cartilage a demanding field of research within the biomedical engineering industry. There is ample information in the literature on the three different types of cartilage: hyaline, fibro, and elastic cartilage. Studies of the composition of articular cartilage concluded that the ECM is made up of collagen, proteoglycans, chondrocytes, and water. Collagen is the main structure in the extracellular space, while proteoglycans are heavily glycosylated proteins with a core protein with one or more covalently attached sugar chains. Chondrocytes are the single existing cell within the articular cartilage and are highly specialized and metabolically active while water is the most abundant component contributing to approximately 80% of the wet weight. Further research into the structure of articular cartilage (AC) found four distinct zones, including superficial, middle, deep, and calcified; all of which have distinct differences in composition. Within these four zones, three regions are present referenced as the territorial, pericellular, and interterritorial region. Initiating signal transduction throughout the loaded cartilage between chondrocytes and the ECM is controlled by the territorial region, which also surrounds the pericellular region. Bundles of large collagen fibers that are randomly oriented and have varying proteoglycans define the interterritorial region. Articular cartilage was found to be biphasic and anisotropic giving it distinctly different tensile and compressive properties. The viscoelastic nature of AC is also governed by its liquid and solid ECM phases leading to creep and stress relaxation behavior. Boundary layer lubrication transitioning to fluid film lubrication yields the very low coefficient of friction required to prevent cartilage deterioration between joints. Mechanical materials available today do not present the combination of elasticity and strength articular cartilage possesses making medical replacements less than ideal.

Finally, medical conditions and treatments were covered including osteoarthritis, microfracture, osteotomy, arthroplasty, stem cell therapy, and hydrogels. Current treatments do provide some alleviation of pain and discomfort; however, high costs and unwanted growth of fibro instead of hyaline cartilage leave room for further improvement in this area of the biomedical industry.

4. Biomechanics of atherosclerosis

Currently, atherosclerosis is the leading cause of deaths in the developed world. Due to the nature of this disease and how prevalent it is in our modern society, it is important for all to understand how this disease manifests and is detected, as well as how it can be prevented. Atherosclerosis is a silent killer and contributes to many deaths throughout the world every year. It develops over a long period of time due to a buildup of LDL on the arterial wall before the body contains it with a mesh of collagen and elastin fibers called a fibrous cap. Based on the volume that the soft plaque occupies and the thickness of the fibrous cap itself, a person can live for years without that plaque being a danger to their health. The issue arises when the necrotic core takes up 40% or more of the volume of the plaque or if the fibrous cap is less than 65 μm thick. If either of these criteria are met, the plaque is at risk of rupturing and creating a thrombus that will kill the victim. Even without rupturing, the stiffening of the arterial wall leads to physical changes in the body's function. These changes include increased shear stress in the blood vessels, changes in systolic and diastolic blood pressures, and increased pressure fluctuations in the arteries. Thankfully in our modern age, atherosclerotic plaques can be detected and treated through a variety of methods. Each method has its advantages and disadvantages. Certain procedures excel at identifying atherosclerosis in certain arteries but struggle in other arteries. For these reasons, multiple procedures can be used together to allow a better picture of the health of a patient's arteries. It is becoming more common to use noninvasive procedures as they become more advanced. In many cases, they are just as effective as invasive procedures at detecting atherosclerosis. As atherosclerosis becomes more common, research on the disease is expanding. Every year, doctors are becoming more equipped to handle atherosclerosis.

5. Knee joint biomechanics

During everyday activities like walking and climbing up stairs, the knee experiences 2.7–4.9 times the body weight. It is a wonder how the knee can withstand so much weight over such a long period of time. Millions of years of evolution and adaptation can be found in the knee as it seems to be built for exactly what humans need it to do. The knee is meant to bear load and help the body move at the same time, and it does just that. The meniscus not only reduces the friction between our femur and tibia but also reduces the stress felt on the tibia. Humans often stand for longer periods of time, and the knee has a mechanism to reduce the amount of force needed to stay standing upright. More recently in the field of biomedical engineering, procedures are being developed to fix the issues caused by trauma or simply by degradation.

6. Muscle biomechanics

Muscle is a complex tissue that is involved in many processes and systems essential to human life. They are used for balance, stability, movement, organ function, lifting objects, and many other things. This tissue needs to be analyzed from the smallest unit all the way up to the macroscale of its involvement in complex functions. Furthermore, certain engineering applications and considerations needs to be covered. Through analysis of fundamental structures and functions, the role of the muscle in the body needs to be explored, and its relevance to human life needs to be discussed. Muscle is a contractile tissue within the body responsible for internal and external locomotion and posture. There are different types of muscle tissue, composed of different types of fibers. Contractions are stimulated by the nervous system and have a specific need for energy depending on their environment. In addition, adaptations to the structure and function will arise when exposed to stimuli such as exercise. Muscle is a complex, versatile tissue which provides humans with the ability to execute a variety of tasks. Both muscles as a complex functional group and single muscle fibers as an individual functional unit have been discussed. The interactions of muscle tissue with the rest of the body perform many different functions that need to be carried out simply and effectively. From the contraction of single muscle fibers all the way to the entire muscle groups working together to complete a movement, the structure and fundamental properties of muscles work cooperatively to fulfill different functions and needs of the body.

7. Vascular grafts

Proper circulation of blood throughout the body is extremely crucial to a long and healthy life. Any infliction that inhibits the normal flow of blood through the vast network of blood vessels poses an extreme risk to patients and can cause numerous potentially fatal conditions depending on their location. Vascular grafts are a surgical method utilized to redirect blood flow from one area to another whether to bypass a clogged or narrowed blood vessel or provide an easy access point for other procedures such as blood dialysis. It is an unfortunate truth that, in their current state, synthetic grafts are not a long-term solution to vascular stenosis and provide only small extensions on expected life spans. With the current state of undesirable compliance mismatch that exists between them and the vessels that they are grafted too, it is clear that there is a lot of room for improvement. Future research on the use of multicomponent synthetic grafts in which multiple materials are used together to better mimic the elastin and collagen mechanical properties of natural arteries and veins has a potential for improving the compliance of synthetic grafts, which ultimately leads to improved patency over time. As for tissue engineering, it is being constantly improved upon every single day. The future of biomedical engineering lays in the replication of human tissues through tissue engineering. The only way to correctly mimic the compliance of a human blood vessel is to use a form of tissue engineering. Many improvements have been made already, with some very promising results as seen in the hybrid scaffold methods as well as the decellularized matrices. It is important to note that the highest potential lies in the assembly processes. These are the processes that will allow for any mechanical property and shape to be designed exactly. The only limiting factor being the excessive amount of time required in order to manufacture these grafts. Finally the research toward sutures and the anastomotic site is also a very key area. With the hypercompliant zone being so detrimental, it is very important to consider alternate forms of sutures to combat the high compliance mismatch of those areas.

In the future it can be said that a form of biocompatible glue or laser or even a combination of both may be the best choice of suture. In order for this to happen a lot of work needs to be done in those areas in order to formulate techniques in which the negatives previously mentioned are mitigated.

8. Transcatheter heart valves

Severe aortic stenosis is the calcification of the aortic heart valve that affects 3% of the world's population over the age of 75. This disease may have a variety of causes, such as age, gender, hyperlipidemia, rheumatic fever, hypertension, heart infection, abnormal stresses, and congenital abnormalities. Due to the extremely invasive nature of open-heart surgery, the mortality rate for older patients is very high, and until 1992 these patients would have had to take the risk of open-heart surgery or have no treatment and left to endure cardiac failure. Henning Rud Andersen invented an alternative surgery to replace the native aortic valve, known as a transcatheter aortic valve replacement (TAVR), also known as a percutaneous aortic valve replacement (PAVR). This surgery is done by using various catheters and medical imaging machines to allow for a replacement valve to be directed up an artery to the diseased native aortic valve. The catheter is most commonly inserted into the iliac artery or femoral artery, but there are other methods surgeons use based on their patient. A sheath is placed in an incision located near the groin to aid in inserting various surgical tools and the replacement valve into the artery. A flexible guide wire is transported to the valve to guide the surgical tools and new valve to the native valve. The native valve is then crushed using a procedure called aortic balloon valvuloplasty. Doctors will often use a method called fast pacing during an aortic balloon valvuloplasty to reduce the pulsatile aortic flow by increasing the heart rate to approximately 200 beats/min or greater. After crushing the native valve, a new bioprosthetic aortic valve is set in place using either a balloon expandable PAV or a self-expanding PAV. Doctors use various medical imaging techniques such as fluoroscopy, aortography, and echocardiography in the procedure to monitor flow characteristics and valve deployment location. This procedure is far less invasive than open-heart surgery and gives a safe alternative for older patients or patients characterized with a high mortality rate to receive a new aortic valve. The percutaneous valve is still a growing technology and is still in its optimizing stage. Issues with the valve include thrombosis (blood clotting), valve migration (due to the valve not being sutured in), stent malposition (due to physician error or valve migration), coronary obstruction, and issues with the catheter-based delivery and valve durability. Research into correcting these issues is essential for further optimizing the current models of the percutaneous heart valve and for minimizing negative inoperative and postoperative implications.

Author details

Hadi Mohammadi
Heart Valve Performance Laboratory, School of Engineering, University of British Columbia, Kelowna, BC, Canada

*Address all correspondence to: hadi.mohammadi@ubc.ca

References

[1] Mohammadi H, Mequanint K, Herzog W. Computational aspects of mechanical modeling of articular cartilage. IMechE, Part H: Journal of Engineering in Medicine. 2013;**227**(4):402-420

[2] Mohammadi H, Mequanint K, Herzog W. Multiscale modeling of O_2 transport in articular cartilage. Biophysical Journal. 2012;**102**(3):592a

[3] Raustin R, Mohammadi H. Towards multiscale modeling of wave propagation in arteries. Journal of Mechanics in Medicine and Biology. 2015;**16**(2):1650027 (1-10)

[4] Mohammadi H, Mequanint K. Prosthetic aortic heart valves: Modeling and design. Medical Engineering & Physics. 2011;**33**(2):131-147

[5] Mohammadi H, Fradet G. Prosthetic aortic heart valves: Review article. Cardiovascular System. 2017;**2**(1). 55p

[6] Haley JP, Mohammadi H, Boughner DR. The effects of hammer pressure on cellular response in a porcine heart valve tissue. Cardiovascular Engineering. 2010;**10**(3):157-162

Section 2

Hemodynamics and Blood

Chapter 2

Biomechanics of Human Blood

Emily Earl and Hadi Mohammadi

Abstract

Hematology is known as the study of blood regarding health and disease which revolves around issues with red blood cells (RBCs), white blood cells (WBCs), platelets, lymph nodes, blood vessels, bone marrow, and the proteins involved in bleeding and clotting. As biomedical engineers, it is especially vital to understand the mechanics of the various components of blood to avoid unwanted results from implantations such as heart valves. A comprehensive review on the biomechanics of blood is discussed in this chapter. We will also discuss that even though human blood is a non-Newtonian fluid, depending on the instance, it can be considered a Newtonian fluid.

Keywords: bloodstream, viscosity, hemodynamics, RBC, multiphase flow, pulsatile flow, non-Newtonian flow

1. Introduction

Blood is known to be one of the connective tissues in the human body as the blood connects every single cell, tissue, and organ in the body together [1]. All necessary substances are transported through the vascular system. The science of blood flow and the mechanics of blood flow is known as hemodynamics. Hemodynamics is a significant element of cardiovascular mechanics and engineering as it simply clarifies the physical laws that direct the bloodstream within the blood vessels [2]. A considerable number of dysfunctions that occur due to cardiovascular diseases and disorders such as hypertension and congestive heart failure are linked to systemic hemodynamics. For example, clinical studies advocate that local wall shear stress rates and forms moderate the location and the advancement of atherosclerotic plaques.

The shear stress and wall shear stress in the bloodstream throughout the entire cardiovascular system is significantly impacted by the physics and mechanics of the blood. In particular, these values play and important role in the design and development of medical devices for cardiovascular applications. In the area of cardiovascular engineering and technology and medical devices, hemodynamics and the mechanics of blood are undisputedly vital to be well understood and considered.

2. Blood cells

The major components of blood are considered to be plasma, RBCs, WBCs, and platelets. The liquid component of the blood, known as plasma, is made up of water, salt, sugar, fat, and protein and is responsible for transportation of blood cells throughout the body. Antibodies, oxygen, waste products, chemical messengers such as hormones and proteins, and clotting proteins are also transported throughout the body within the blood. WBCs which are responsible for protection of the body against infection are much fewer in number than RBCs (1 to ~800) [3]. Platelets are not considered as cells but rather a small fragment of cells. Platelets mainly participate in the clotting process, also known as coagulation. They gather at the site of injury and stick to the lining of the injured vessel, forming a frame-like structure on which blood coagulation can take place. This process is also known as the formation of a fibrin clot in which the wound is covered and the leakage of blood is stopped. Fibrin also contributes to the structure of a scaffold upon which new tissue can grow and form, a step known as the healing process.

Red blood cells: RBCs are about 40–45% of the blood volume. Their shape is like a biconcave disk with a flattened center [4]. RBCs are made up of a compound known as hemoglobin which is a protein for carrying oxygen. Hemoglobin consists of two alpha subunits and two beta units. Each subunit holds a heme group and each heme holds a Fe^{2+} ion which can bind to an O_2 molecule. Oxyhemoglobin is essentially a hemoglobin that has molecules for bonding to oxygen molecules. RBCs undergo two major states while circulating through the body, oxygenated and deoxygenated. Oxygenated cells are bright red color and contain large quantities of oxyhemoglobin. They circulate through the body to deliver oxygen to the body tissues. When an RBC reaches the intended tissue, oxygen molecules are removed from hemoglobin. The first two O_2 molecules are easier to remove than last two, which cause a gradient of release. Deoxygenated cells have less oxyhemoglobin existing in the hemoglobin compound, however blood is never actually deoxygenated as not all oxygen is removed [5].

Blood disorders: Some common blood disorders include anemia, malaria, and cancer. Anemia occurs when the number of red blood cells is comparatively low. Common causes of anemia include iron deficiency, B_{12} deficiency, chronic diseases of the kidney or bones, and red blood cell destruction due to shearing forces. Malaria is a mosquito-borne infectious disease. Minor symptoms include fever, fatigue, vomiting, and headaches, however severe symptoms can include seizures, coma, and possibly death. The main cancer associated with blood is leukemia which begins in the bone marrow and results in high numbers of abnormal white blood cells. These abnormal white blood cells, known as blasts, are not fully developed and cannot function properly. This causes symptoms such as bleeding and bruising, fatigue, fever, and an increased risk of infection. Certain disorders such as thalassemia and leukemia have varying types for which symptoms and ideal treatment varies.

Thalassemia: Thalassemia is an inherited blood disorder where the hemoglobin produced by the body is abnormal and does not function properly. Thalassemia develops due to a genetic mutation in one of the genes involved in the production of hemoglobin. The disorder leads to the destruction of RBCs which results in anemia [6].

There are three main forms of thalassemia known as alpha thalassemia, beta thalassemia, and thalassemia minor [6]. Thalassemia minor is the less severe form, whereas alpha thalassemia and beta thalassemia are more serious conditions. Alpha thalassemia occurs when at least one of the alpha globin genes has a mutation, while beta thalassemia occurs due to mutations in the beta globin genes. Alpha and beta thalassemia also have two subtypes. Beta thalassemia has the subtypes major and intermedia. Beta thalassemia major is the more severe form of this disease and is generally diagnosed early on when the child is in infancy. Patients with beta thalassemia major have a complete lack of beta globin genes and experience the most severe symptoms. In addition to severe anemia which can be life-threatening, other symptoms may include paleness, poor appetite, jaundice, and enlarged organs [6]. Beta thalassemia intermedia is the less severe form of beta thalassemia and arises due mutations in the beta globin genes. Unlike beta thalassemia major, the genes are present within the DNA just not in their normal form.

Alpha thalassemia has the two subtypes hemoglobin H and hydrops fetalis. Hemoglobin H is developed when a patient is missing up to three alpha globin genes or has mutations in up to these alpha globin genes. Complications from this disease can cause bone issues where the cheeks, forehead, and jaw overgrow. Individuals may also experience jaundice, malnourishment, and an extremely enlarged spleen [6]. Alpha hydrops fetalis thalassemia is an extremely severe form of the disease and occurs in the developing fetus. This condition develops when all four alpha globin genes are altered or missing [6]. Due to such early development babies with this form of thalassemia are usually stillborn or die shortly after birth.

The only treatment for patients with thalassemia is regular blood transfusions every 2–4 weeks. This causes the patients to have excess iron which can cause iron overload in the body and lead to dangerous side effects. Due to this, patients also need to take drugs called iron-chelating agents that bind to excess iron and help the body remove it from their systems [7, 8].

Leukemia: In leukemia the DNA of immature blood cells, most commonly, WBCs is damaged which causes WBCs to grow and replicate continuously. Unlike healthy WBCs, these abnormal blood cells continue to accumulate in the bloodstream, forcing out healthy cells. As the damaged WBCs grow, they start to affect the normal functions of healthy WBCs by filling up large amounts of space within the blood. Individuals with leukemia generally suffer from poor blood clotting, anemia, and weak immune systems. Other symptoms that can be experienced are nausea, fever, chills, night sweats, flu-like symptoms, weight loss, bone pain, and tiredness [8].

Leukemia is split into two sets of types classified as acute or chronic and lymphocytic or myelogenous. Chronic leukemia is a rapidly moving form of cancer, whereas acute leukemia progresses significantly slower. Once divided into either acute or chronic, leukemia is then subdivided by the type of affected blood cell. Lymphocytic leukemia describes cancerous cells affecting the bone marrow that makes lymphocytes. Whereas myelogenous leukemia covers cancers that occur in the bone marrow that produce other types of white blood cells, platelets, and RBCs.

Treatments for leukemia vary depending on the classification of the cancer as well as the age and general health of the patient. Treatments for acute leukemia should be started as soon as possible due to the aggressive nature of the cancer and include chemotherapy and bone marrow transplants. Chronic leukemia is treated differently depending on the stage. The types of treatment include targeted therapy, interferons, chemotherapy, radiation therapy, and stem cell transplants [8].

The life cycle of RBCs: RBCs are produced in the bone marrow through a process known as erythropoiesis. The production of RBCs involves erythropoietin, monosaccharides, lipids, vitamin B_{12}, amino acids, folic acid, and iron. RBCs are released into the bloodstream once they are developed and have a lifespan of 120 day after which they expire due to mechanical or structural damage. Dead RBCs are then removed from the bloodstream through the spleen, liver, and bone marrow. The dead cells are crushed into their main components known as heme, comprised of iron and bilirubin, and globin, comprised of amino acids. Amino acids and iron are reused by bone marrow, whereas the bilirubin is removed through feces and urine [9].

RBC configuration: RBCs have a discrete biconcave shape, similar to a disk (**Figure 1**). Normal cells are 7.5–8.0 μm in diameter and ~2.0 μm in height, however RBCs must adapt their shape in order to pass through capillaries as some capillaries are only ~3 μm wide [9]. This adaptation means RBCs feel significant passive deformation through their 120-day lifespan. The properties of a RBC must be physically and mechanically stable so that to resist disintegration and the mechanical properties have to do significantly with their deformation in terms of the bending, shear, area expansion moduli, and relaxation times.

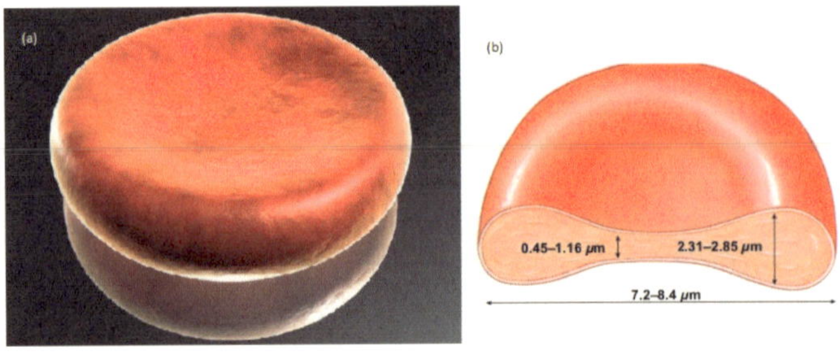

Figure 1.
The geometry of human RBCs. (a) The normal biconcave-shaped RBC cut in half along the y-z plane. (b) A RBC with the cytoplasm [10].

Microfluidic channels are implemented to simulate human capillaries and study RBC deformability. Microfluidic channel are formed from stationary DPD particles and are filled with fluid particles containing RBCs. The microfluidic channels are comprised of two wide channels on each other edge and a cuboid channel in between. The wide channels are normally 20.0 μm wide and 3.0 μm high whereas the cuboid channel is 4.0 μm wide, 3.0 μm high, and 30.0 μm long [9]. Adaptive boundary conditions for fluid DPD particles are used to control density fluctuations. An individual RBC undergoes a continuous and severe transition from its normal biconcave shape to an ellipsoidal shape. This deformation includes elongation in the flow direction (longitudinal axis), and shortening in the cross-flow direction (transverse axis). The RBC enters into the narrow channel by undergoing these deformations. Once the entire RBC enters the constriction, it deforms further to pass through the microfluidic channel.

RBC deformation: Three distinct cellular components contribute to RBC's deformability—cell geometry, cytoplasm viscosity, and membrane elasticity.

Cell geometry: The geometry of a RBC determines the ratio of cell surface area to cell volume. The deformation of a cell is directly related to the surface area to cell volume ratio. Therefore, cells that achieve a higher value of surface area to volume ratio can deform with more ease.

Cytoplasm viscosity: The cytoplasm viscosity is regulated by the mean corpuscular hemoglobin concentration (MCHC). This suggests that the viscosity is directly related to alterations in cell volume. Therefore, cells with greater volumes have a higher cytoplasm viscosity and will deform with more difficulty.

Membrane elasticity: The membrane of RBCs consists of a lipid bilayer supported by an attached spectrin-based cytoskeleton. The resistance of the lipid bilayer to bending elasticity is controlled by the bending rigidity ($\mathbf{k}_c$). The spectrin network's resistance to shear strain is characterized by the in-plane shear modulus, ($\boldsymbol{\mu}_s$). The deformability of the membrane, along with the mechanical stability of the cell, can be attributed to the elastic modulus, bending modulus, and yield stress.

Membrane simulations: RBC membrane properties can reveal the complex behavior that takes place within the membrane when it deforms. Twisting torque cytometry (TTC) is used to simulate membrane rheology and can obtain RBC membrane properties such as yield stress, shear thinning, and viscoelasticity. A microbead is bonded to the surface of a cell membrane and a magnetic twisting cytometry applies both static and oscillating magnetic field. The wall adhesion is simulated by keeping 15% of vertices stationary on the bottom of the lipid bilayer component of the RBC membrane. Microbead adhesion is simulated by including several RBC vertices in the lipid bilayer component near the bottom of the microbead in its rigid motion.

The complex elastic moduli of a RBC can be computed from the phase angle between the storage and loss moduli as such:

$$g' = \frac{\Delta T}{\Delta d} cos\varphi \tag{1}$$

$$g'' = \frac{\Delta T}{\Delta d} sin\varphi \tag{2}$$

Here, φ is the phase, g' is the two-dimensional storage, g'' is the loss modulus, T is the torque, and d is the microbead displacement amplitudes.

3. Hemodynamics and flow types

Blood circulation: Blood flow in the circulatory system is determined by the pulsing drive that is developed from the heart, the individual mechanical and flow properties of the fluid, and the structure and mechanical properties of blood vessels. These factors combined at appropriate levels ensure that the cells of the body receive adequate amounts of oxygen as well as maintain waste management.

Flow pulse development: The main function of the heart is to circulate blood throughout the human body. It is composed of four chambers: two chambers known as ventricles on the lower half of the heart and two chambers known as atria composing the upper section as shown in **Figure 3**. Upon the propagation of bioelectricity through these different components of the heart, contraction of each chamber occurs, moving blood throughout the body in a system known as the cardiac cycle [11]. The cardiac cycle can be easily separated into two main time events: systole and diastole. These two events refer to the action of either the heart pumping blood into the circulatory system or receiving blood from the venous system. In addition to these events, other factors of the blood such as velocity are initialized from the cardiac output of the heart.

Systole and diastole: Systole is when the pressure in the circulatory system is the highest due to the force of the heart that is used to pumping blood into the aorta and pulmonary artery, whereas during diastole the blood is moved into the heart due to a pressure difference between the vena cava and the right atrium, and pressure is lowest [12]. These periodic variations in pressure is what causes blood flow to be considered “pulsatile” [13]. This pulsatile action is what makes blood flow unable to be effectively modeled by standard flow models unless specific assumptions are applied.

Cardiac output: The amount of blood that flows out of the heart in 1 minute is known as the cardiac output and varies dependent on the weight of an individual. Standard values of cardiac output are within the range 4.0–8.0 L/min [13]. Cardiac output is dependent on four main components: heart rate, contractility, preload, and afterload.

The heart rate is directly proportional to the velocity of the blood moving throughout the body because under normal circumstances the blood maintains a constant volume. As heart rate increases, the velocity increases, which affects the viscosity and turbulent effects of the flow. A similar relationship is seen with contractility as the greater the force the heart initially enacts while emptying the left ventricle, the larger the pressure will be that is pushing the blood from the heart, increasing the initial blood velocity. Preload is the degree of myocardial extension prior to shortening which maintains a direct relationship with cardiac output [3]. Afterload is the force that the ventricle must overcome in order to push the blood into the system of blood vessels. These components combined affect the initial velocity, pressure, and forces applied on the blood flow. However, this system is only effective if the components of the heart are working properly and can differ if there are defects present in the heart.

Heart murmurs: A heart murmur is a sound that is developed in the heart that occurs due to the presence of turbulent flow near the heart valves [14]. Heart murmurs can be classified into two main types: innocent and abnormal murmurs. These murmurs can then be categorized based on if they occur during systole or diastole and by what type of flow characteristic they possess, namely, regurgitation or ejection [15]. Due to the fact that heart murmurs occur due to turbulent flow, they have a tendency to be increased in those who are diagnosed with anemia (due to the decreased hematocrit) and those with heart valve defects. In addition to these two cases, anything that causes irregular or disturbed flow has the potential to cause increased turbulence in the flow and therefore increase the possibility of heart murmurs. Examples of such include heart valve replacements which introduce new stresses and area contact points in the flow and heart valve infections which cause inflammation [15].

Blood circulation begins by the heart pumping deoxygenated RBCs to the lungs which are then oxygenated and released back to the heart through the pulmonary circuit. These oxygenated RBCs are then pumped through the systemic circuit to deliver oxygen to various tissues. The RBCs become deoxygenated after releasing and depositing oxygen within the tissues and travel back to the heart through the system circuit to repeat the cycle.

The flow of fluid within the circulatory system is dependent on a variety of factors but can be characterized by considering the laminar and turbulent properties of the flow. In addition to the laminar and turbulent properties of the flow, it is also important to consider the motion of the suspended particles within the heterogenous fluid allowing it to cohere to adequate blood flow needs.

Laminar and turbulent blood flow: In homogeneous fluids, flows are laminar up to a Reynolds number of roughly 2300 and become turbulent at a Reynolds number of 4000. This logic cannot be applied to the flow of blood as blood is not a homogenous fluid and blood vessels are not perfectly cylindrical and possess viscoelastic properties. Though directly corresponding to Reynolds numbers will not accurately represent the type blood flow, it is generally considered that the possibility of turbulence will increase as the Reynolds number increases, regardless of the precise critical Reynolds number values for transition. Adhering to this logic, as seen in Eq. (3), the possibility for turbulent flow will increase as the velocity increases, the diameter increases, the density increases, or as the viscosity decreases:

$$\mathfrak{R} = \frac{\rho N D^2}{\mu} = \frac{\rho V D}{\mu} \tag{3}$$

As a flow develops into turbulence unsteady vortices appear and interact with each other leading to the development of eddy currents, small currents where the flow differs from that of the general flow. Turbulence occurs naturally in locations of the circulatory system where the Reynolds numbers are comparatively elevated such as the ventricles and ascending aorta. In addition to this, turbulence can also be initiated due to branches or curves in the flow, irregularities due to surgical implants, and improper function of circulatory valves [16]. Under diseased or abnormal conditions, other segments in the circulatory system can experience turbulent flow which can have negative effects on epithelial function [16].

In individuals with conditions affecting the viscosity of their blood, such as anemia, due to the decreased hematocrit in the blood the opportunity for one's blood to enter turbulence is increased [5]. Other individuals experience increased turbulence opportunity due to a foreign object being placed in the circulatory system such as a replaced heart valve, as turbulence is developed from an increased contact area between the blood flow and the valve. Due to the chaotic nature of the flow, turbulent flows require more energy to properly travel throughout the system as much of the energy is lost due to misaligned flows and eddy currents. Even though turbulent flows occur in the circulatory system, RBCs go through a series of motions and deformations which help sustain the efficiency of blood flow.

Movement of RBCs: Dependent on shear rate, RBCs can move throughout the circulatory system in one of three manners known as tumbling, swinging, and tank-treading [17]. At the lowest shear rates RBCs have the tendency to move in a pattern known as tumbling (**Figure 2**). This is where the RBCs spin completely around their axis and maintain little to no deformation. As shear stress is increased RBCs experience a transition motion known as swinging where they experience quasi-deformation and their rotation abilities alter from 180° at tumbling to a range of 5–26° while swinging [5]. Following this, RBC motion develops completely into tank-treading motion which is defined by large amounts of RBC deformation and quasi-steady motion. During tank-treading, RBCs explore a small volume of the flow, leading to less collisions and disturbances, which decreases the viscosity of the fluid. This phenomenon coheres with the fact that as shear rate increases, the viscosity of the blood decreases.

Viscosity: The viscosity of the blood is due to the internal friction between the flow, incorporating the effects of the suspended particles present in the blood, inclusive of RBCs, WBCs, and platelets. As this internal friction increases, more force is required from the heart in order for it to maintain the desired cardiac output of the blood in the circulatory system. This requires a heightened contractility from the muscles of the heart which can result in the fatigue of the heart and, in major cases, heart collapse [19]. The opposite case where there is a lack of proper internal friction in blood flow will cause a decrease in the ability of one's blood to clot, which imposes risk when blood vessels are damaged and the blood continues to flow out of the site of damage for a prolonged period of time.

The viscosity of blood is dependent on many factors such as the properties of blood plasma, the hematocrit levels, and the individual mechanical properties and influence of the suspended particles in the flow; however, this is inherently dependent on whether blood is considered as a Newtonian or non-Newtonian fluid. The true nature of blood is that it exhibits non-Newtonian properties under specific

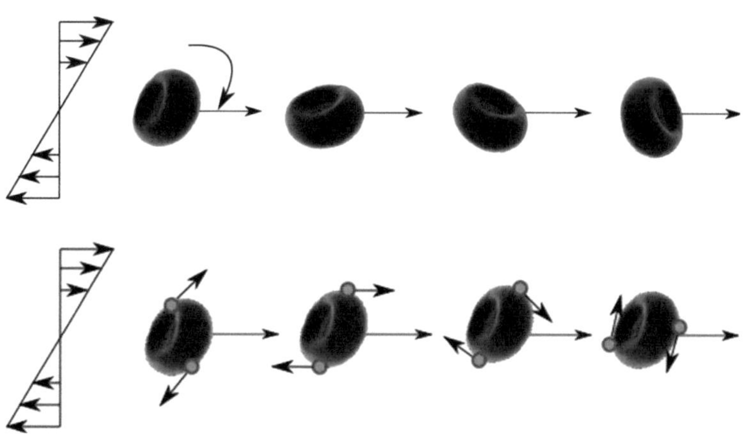

Figure 2.
Two conceptual motions are considered for RBCs, (1) linear shear field, the solid-like, also known as tumbling motion (top), and the vesicular, also known as tank treading motion (bottom) [18].

conditions; however, these conditions arise at very few locations throughout the circulatory system.

Newtonian and non-Newtonian conditions: Blood possesses non-Newtonian properties when the shear rate is above 100 s^{-1} [20] and follows shear-thinning effects. High shear rates occur in the capillaries and the larger arteries coming directly from the heart because the shear rate of a fluid through a vessel increases with increasing velocity and decreasing diameter. Due to the fact that the main non-Newtonian properties of blood occur in small diameter vessels, it is argued that the non-Newtonian effects that occur within the largest arteries can be ignored [21, 22]. In blood vessels that possess a diameter in the more medial range or where there is decreased velocity, such as veins and arterioles, these non-Newtonian factors have minute effects on the properties of the flow, causing them to be neglected for these areas as well:

$$\acute{\gamma} = \frac{4Q}{\pi r^3} \tag{4}$$

Eq. (4) is not an accurate representation of proper blood flow as blood flow is subject to fluctuations based on the viscoelastic properties of the vessel walls, the alternating pressures from systolic and diastolic, and when present the non-Newtonian properties of the blood itself; however, it is used here to outline the relationship between the shear rate and vessel size and diameter.

Considering the above, the non-Newtonian effects of blood are only active within a small portion of overall blood flow and acquire more importance when blood flow at those selected areas are specifically studied. When being considered as a Newtonian fluid, the viscosity of blood in addition to being impacted from the vessel size and blood velocity is also affected by the blood plasma and the concentration of suspended particles in the flow.

Factors of blood as a Newtonian fluid: The blood plasma is mainly composed of water (roughly 91% by volume) proteins, hormones, and glucose and acts as a Newtonian fluid with standard values between 1.1 and 1.3 mPas at the human body temperature of 37°C [9]. Blood plasma accounts for approximately 55% of the volume of the blood in the body and due to the incredibly high water content in the substance, the viscosity of plasma is highly affected by the hydration levels of the individual. As a human becomes dehydrated this percentage decreases and the blood becomes more viscous [23]. In addition to hydration levels, blood plasma viscosity is also directly affected by the amount of proteins and lipids in the blood post consumption. The higher the concentration of these elements, the more viscous the plasma will become [24].

In addition to the direct effects on plasma, the shear in the flow is also affected by the amount of suspended particles in the flow. In heterogeneous fluids where particulates are present, these particulates alter the velocity profile of the fluid due to the increased shear at the fluid particle interface. As aside from plasma, the majority of the remaining volume of the blood is composed of RBCs, RBCs are the particles that impose this effect to the flow in the greatest magnitude. Aside from the direct viscosity of the plasma itself, plasma also affects the viscosity of the blood by the housing of certain proteins such as fibrinogen that cause aggregation in the suspended particles [25].

Factors of blood as a non-Newtonian fluid: The percentage comparison of the volume of blood cells to the total volume of blood is known as hematocrit and is the main factor contributing to the viscosity of the blood. This is because the blood's ability to flow is highly dominated the ease of movement of RBCs. At high shear rates the deformability of RBCs is what effectively determines the viscosity of the fluid; however, at low shear rates the viscosity is controlled by the unique property of RBCs to aggregate [26, 27].

Physical capabilities and tendencies of RBCs: The deformability of RBCs is controlled by three main factors: the relatively high surface area to volume ratio due to RBCs enucleated nature, the viscosity of the cytoplasm, and the viscoelastic properties of the cell membrane [28]. The viscoelastic properties of the cell membrane are dominated by three moduli known as the shear elastic modulus, the area compressibility modulus (κ), and the bending modulus. The definitions of the previous as well as determined experimental values for healthy RBCs are denoted in **Table 1**. The bending modulus (E_K) is calculated as such:

$$E_K = 2KH^2 \tag{5}$$

$$H = \frac{1}{2}\left(\frac{1}{R_1} - \frac{1}{R_2}\right) \tag{6}$$

During the same experiment as the calculation for the elastic moduli, the cytoplasmic viscosity was tested as well, producing a value with an average that is approximately six times greater than the viscosity of plasma.

This viscosity is important because it outlines how quickly the cell can reshape itself. Similar to the viscosity of plasma, this is dependent on the hydration levels of the individual in which the blood is present [30]. Deformability of RBCs is relevant in locations of high shear rates such as the capillaries because, in order to maintain proper blood flow they must adhere to the vessel to sustain motion. To do this effectively, the RBCs fully elongate into ellipsoids and align with the flow, reducing the possibility of collisions, decreasing the overall viscosity.

At low shear rates, RBCs have the tendency to aggregate together, most commonly into stacks called Rouleaux. It is suggested that this specific formation occurs due to the incredibly high surface area RBCs possess. This combining of RBCs severely increases the frictional resistance between flow streamlines, increasing the viscosity of the fluid. However, as seen in **Figure 3** at high shear forces this tendency is overruled and the blood cells separate and align in the direction of the flow [31].

Aside from RBCs, other suspended particles such as platelets and WBCs are present in the blood which also maintain aggregative properties; however due to the fact that they compose roughly 1/800th and 1/600th of the volume of the blood, respectively, they are often not considered a vital part of the viscosity of the blood [31–34].

Flow effects on RBCs: As blood moves from a large vessel to a vessel less than 0.3 mm [35] the RBCs realign to the center of the vessel. Due to this, the velocity of the centric RBCs is increased relative to the layer of plasma present at the wall of the vessel and the RBCs leave the vessel at a faster rate at which they enter them.

Modulus type	Definition [29]	Tested value [4]
Shear elastic modulus	The ratio of shear stress to shear strain	5.5 +/− 3.3 (μN/m)
Area compressibility modulus	The energy per unit area required to uniformly stretch an interface to produce an area change according to Hooke's law	399 +/− 110 (mN/m)
Bending modulus	The energy per unit area required to produce a mean curvature (H) according to Eqs. (3) and (4)	1.15 +/− 0.9 ($\times 10^{-19}$Nm)

Table 1.
Viscoelastic factors for RBCs.

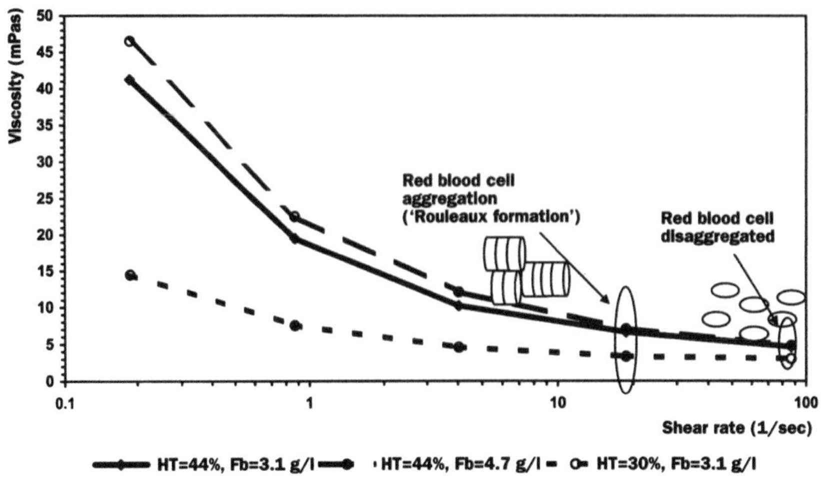

Figure 3.
Shear rate to RBC aggregation comparison [31].

This causes the hematocrit level to decrease through the vessel, also known as the Fahraeus effect. This causes another effect known as the Fahraeus-Lindkvist effect which states that the viscosity of the blood decreases as the vessel size decreases. Though as viscosity in extremely small vessels is affected by the deformability of RBCs as discussed earlier, the increase in velocity of the RBCs increases the velocity of the entire flow, respectively, causing the viscosity to decrease.

Hemostasis: Hemostasis is the process by which bleeding is stopped and it can be broken up into three main steps: vasoconstriction, platelet activation, and blood clot formation [36]. The clotting process begins by the blood vessel contracting in order to reduce the flow of blood into the injured vessel. After a rupture, tissue factor is released which causes platelets to aggregate to each other and the walls of the vessel. As platelets become sticky, they help to impede the flow of blood through the rupture. In addition chemicals are released from small sacs inside the platelets called granules [37] that attract more cells to the site and further the clumping of platelets creating a platelet plug. On the surface of these activated platelets, many clotting factors work together in a series of complex chemical reactions known as the coagulation cascade which results in the formation of a fibrin clot [38].

Blood diluters: Blood diluters are commonly used in the medical industry for a variety of reasons. The main reason that a doctor may prescribe a blood thinner to a patient is if the patient has a high risk of blood clots, which can cause organ damage or in some cases death. Blood thinners are commonly used for patients suffering from heart disease, poor blood circulation, abnormal heartbeat, and congenital heart defects [6, 8, 39–41]. There are two types of blood thinners, anticoagulants and antiplatelets. Anticoagulants inhibit the coagulation cascade, whereas antiplatelets prevent platelets from aggregating. Antiplatelet drugs are commonly issued for patients with heart disease or have had prior heart attacks and anticoagulant drugs are used before open heart surgery on heart valves or congenital heart defects [41]. Using blood thinners inhibits vital aspects of the human body which can cause many side effects. Patients can suffer from increased bruising, red- or pink-colored urine, bloody stools, increased bleeding during menstrual period, purple toes, and blackish areas in their fingers, toes, or feet [9]. Common anticoagulants include heparin and warfarin. Well-known antiplatelet drugs are clopidogrel and ticagrelor [42].

Viscoelasticity of blood: Viscoelasticity is the property of materials that exhibit both viscous and elastic characteristics when undergoing deformation [43–49].

Viscous materials resist shear flow and strain linearly with time when a stress is applied, while elastic materials strain when stretched and quickly return to their original shape after the stress is removed. The blood is viewed most commonly as a viscous fluid, but due to RBCs ability to store elastic energy, it actually displays viscoelastic properties. Elastic energy is most commonly stored in RBCs as a result of the heart pumping the blood.

The factors contributing to the viscoelastic properties of blood are the plasma viscosity, plasma composition, temperature, shear rate, and the level of hematocrit. When conducting experiments on blood to test the viscoelasticity properties there are two main control factors, the level of hematocrit and the temperature.

Hematocrit is a significant factor as RBCs are the main reason for the elastic properties of blood. **Figure 4** shows the level of viscoelasticity with respect to the amount of RBCs present in the blood. It is easy to discern from the graph that as the amount of RBCs present in the blood increases, the viscoelasticity properties of the blood increase as well.

It is important for scientists to properly recreate the same environment experienced within the human body to achieve accurate results. As shown in **Figure 5**, the temperature affects the levels of viscosity and elasticity within the blood.

Blood clot factors: In stagnant flow regions or where the blood flow moves very slowly, the risk for blood to clot increases [50]. This occurs due to a high exposure time of RBCs to large variation in shear stress. It has been shown that the pulsatile flow is significant in the regulation of the stagnation areas regarding blood clot formation [50, 51]. In addition, blood clotting is known to occur because of both the jet velocity and turbulent shear stress where Re number is high in the stagnation region [52]. Factors that are said to be in charge of triggering blood clot formation are listed in **Table 2**.

Numerical models for blood clot formation: The process of blood clotting begins by activated platelets which aggregate with a damaged blood element. It is well known that the level of platelet activation and blood cell damage are significantly impacted by the magnitude and duration of the applied shear stress, known as residual time [59].

There have been a few models developed based on the measured residual time and the amount of shear stresses as outlined in **Table 3**.

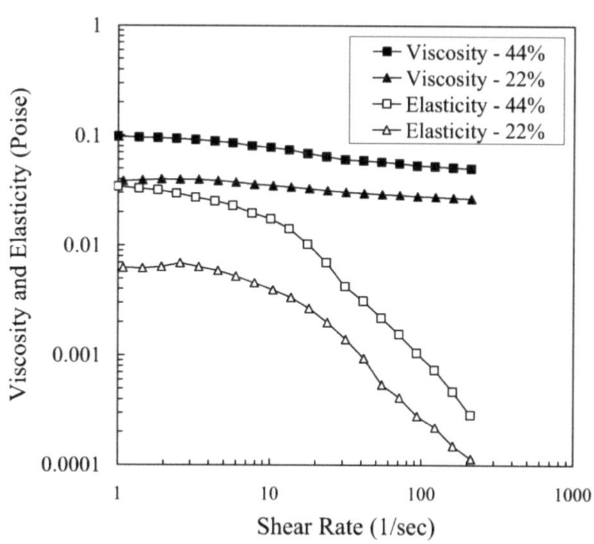

Figure 4.
Viscosity and elasticity measured at 2 Hz, 22°C, and at a shear rate of 10/s rms [43].

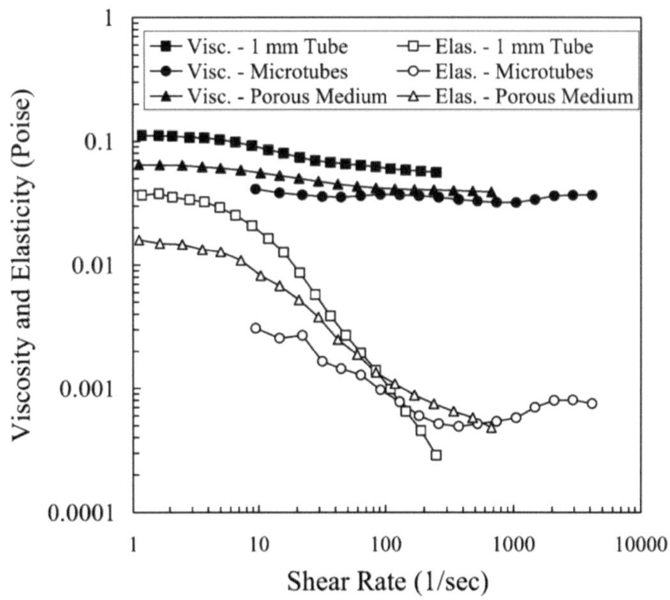

Figure 5.
Viscoelastic properties of blood with respect to shear rate. Hematocrit is set to %45 and measurements are performed at 2 Hz in three arrangements: cylindrical tube of diameter of 0.10 cm and length of 6 cm, in a microtube with a diameter of 0.005 cm and a length of 0.120 cm and porous medium with an equivalent diameter of 0.00812 cm and a length of 0.165 cm.

Factor	Triggering criteria for blood clots
Cavitation	Water hammer and squeeze flow
Reynolds shear stress	>>200 dynes/cm^2 [54]
Cardiac output	Slow movement of leaflet
Stagnant flow	If occurs adjacent to the valves, it could promote the deposition of damaged blood elements, leading to thrombus formation on the prosthesis [55]
Vortex shedding	Yields repeated vortex pairing within the wake, which is responsible for the formation of larger platelet aggregates [56]
Recirculation	Allows many platelets to be trapped [57]
Pressure drop	A larger pressure drop means that the heart with the MHV prosthesis has to work harder [58], thereby reducing cardiac output

Table 2.
Blood clot factors [53].

Model	Expression	
Linear damage accumulation/BDI	$\sum_{t=0}^{t=end}(\tau \times \Delta t)$dynes/$cm^2$	[60]
Platelet activation state (PAS)	Nondimensional level of platelet activation within the interval of [0, 1], in which 0 and 1 correspond to nonactivated and fully activated platelets, respectively	[61]
Power-law model	$C\tau^{\alpha}\tau^{\beta}$	[62]
λ_2 criterion	$\lambda_2 = 20.0s^{-2}$ is responsible for blood clot formation	[63, 64]
Adhesion model	$S \leq S_{th}$, where S_{th} is shear rate threshold, taken as 100	[65]

Table 3.
Available models for the estimation of blood clot formation and threshold [53].

4. Conclusion

Biological processes are amazing in their complexity and optimization. The blood, being no exception, is extremely evolved and adapted to the different scenarios necessary to maintain life. Consisting of plasma, WBCs, platelets, and RBCs, it is able to transport vital molecules around the body including oxygen, and clot in case of injury. Since RBCs make up approximately half the volume of blood, blood flow mechanics are largely related to the properties of red blood cells defined under a soft solid. Large deformability is essential in the life cycle and function of red blood cells as capillaries are extremely small. Blood clotting is a very important function of blood, in which platelets are the main contributors to the clotting process. When platelets come into contact with damaged tissue, the platelets activate and construct a web to coagulate blood. Due to vWF's shear dependent binding, under high shear stresses platelets can be bound together and form clots without activating.

Author details

Emily Earl and Hadi Mohammadi*
The Heart Valve Performance Laboratory, School of Engineering, Faculty of Applied Science, University of British Columbia, Kelowna, BC, Canada

*Address all correspondence to: hadi.mohammadi@ubc.ca

References

[1] Mohammadi H, Mequanint K. Prosthetic aortic heart valves: Modeling and design. Medical Engineering & Physics. 2011;**33**(2):131-147

[2] Mohammadi H, Fradet G. Prosthetic aortic heart valves. In: Cardiovascular Systems. Vol. 5(2). 2017

[3] Vincent JL. (2008). Understanding cardiac output," Critical Care, vol. 12, no. 4, p. 174

[4] Chiu JJ, Chien S. Effects of disturbed flow on vascular endothelium: Pathophysiological basis and clinical perspectives. Physiological Reviews. 2011;**91**(1):327-387

[5] Dupire J, Socol M, Viallat A. Full dynamics of a RBC in shear flow. Proceedings of the National Academy of Sciences of the United States of America. 2012;**109**(51):20808-20813

[6] Emily E, Mohammadi H. Proposed synthetic tissues that replace human cadavers for training. Medical and Surgical Education - Past, Present and Future, Chapter 6. 2018. ISBN 978-953-51-5700-7

[7] Earl E, Mohammadi H. Engineering aspects of human blood. Biomedical Engineering Current Research (Review Article). 2018;**1**(2):4-10

[8] Earl E, Mohammadi H. Improving finite element results in modeling heart valve mechanics, IMECH Part H. Journal of Engineering in Medicine. 2018;**232**(7):718-725

[9] Kesmarky G, Kenyeres P, Rabai M, Toth K. Plasma viscosity: A forgotten variable. Clinical Hemorheology and Microcirculation. 2008;**39**(1-4):243-246

[10] Chee CY, Lee HP, Lu C. Using 3D fluid–structure interaction model to analyse the biomechanical properties of erythrocyte. Physics Letters A. 2008;**372**(9):357-1362

[11] Mohammadi H, Klassen RJ, Wan WK. A finite element model on effects of impact load and cavitation on fatigue crack propagation in mechanical bileaflet aortic heart valve, IMECH, Part H. Journal of Engineering in Medicine. 2008;(7):1115-1125

[12] Millon LE, Mohammadi H, Wan WK. Anisotropic polyvinyl alcohol (pva) for cardiovascular applications. Journal of Biomedical Materials Research. 2006;**79**B(2):305-311

[13] Sabbah HN, Stein PD. Turbulent blood flow in humans its primary role in the production of ejection murmurs. Circulation Research. 1976;**38**(6):513-525

[14] Mohammadi H, Herzog W. A novel model for diffusion based released kinetics using an inverse numerical method. Journal of Medical Engineering & Physics. 2011;**33**(8):893-899

[15] Mohammadi H, Mequanint K, Herzog W. Computational aspects of mechanical modeling of articular cartilage, IMECH, Part H. Journal of Engineering in Medicine. 2013;**227**(4):402-420

[16] Kruger T, Gross M, Raabe D, Varnik F. Crossover from Tumbling to Tank-Treading-Like Motion in Dense Simulated Suspensions of RBCs. London, United Kingdom: Centre for Computational Science; 2013

[17] Kumar D, Vinoth R, Raviraj A, Vijay Shankar CS. Non-Newtonian and Newtonian blood flow in human aorta: A transient analysis. Biomedical Research. 2017;**28**(7):3194-3203

[18] Melchionna S. A model for red blood cells in simulations

of large-scale blood flows. Macromolecular Theory and Simulations. 2011;**20**(7):548-561

[19] Janoschek F, Toschi F, Harting J. Physical Review E. 2010;**82**:056710

[20] Haley JP, Mohammadi H, Boughner DR. The effects of hammer pressure on cellular response in a porcine heart valve tissue. Cardiovascular Engineering. 2010;**10**(3):157-162

[21] Perktold K, Peter R, Resch M. Pulsatile non-Newtonian blood flow simulation through a bifurcation with an aneurism. Biorheology. 1989;**26**:1011-1030

[22] Ballyk PD, Steinman DA, Ethier CR. Simulation of non-Newtonian blood flow in an end-to-side anastomosis. Biorheology. 1994;**31**:565-586

[23] Jeukendrup A, Gleeson M. Dehydration and its Effects on Performance. Human-Kinetics. 2017. 30-04-2009 [Online]. Available from: http://www.humankinetics.com/excerpts/excerpts/dehydration-and-its-effects-on-performance [Accessed: 05-15-2017]

[24] Dintenfass L. Blood Viscosity. Springer Science & Business Media; 1985

[25] Zareh M, Fradet G, Naser B, Mohammadi H. Are two-dimensional images sufficient to assess the atherosclerotic plaque vulnerability? A viscoelastic and anisotropic finite element model. Journal of Cardiovascular System. 2015;**3**(3):1-11

[26] George JN. Platelets. Platelets on the Web. 2015 [Online]. Available from: 04-06-2015

[27] Simmonds MJ, Meiselman HJ, Baskurt OK. Blood rheology and aging. Journal of Geriatric Cardiology. 2013;**10**(3):291-301

[28] Mouritsen OG. Life—As a Matter of Fat: The Emerging Science of Lipidomics. Springer Science & Business Media; 2006

[29] Tomaiuolo G. Biomechanical properties of RBCs in health and disease towards microfluidics. Biomicrofluidics. 2014, 2014;**8**(5):051501

[30] Effros RM, Chang RS, Silverman P. Effect of osmolality on RBC viscosity and transit through the lung. Journal of Applied Physiology. 1977;**42**(6):941-945

[31] Pop GAM et al. The clinical significance of whole blood viscosity in (cardio)vascular medicine. Netherlands Heart Journal. 2002;**10**(12):512-516

[32] Bessonov N, Sequiera A, Simakov S, Vassilevskii Y, Volpert V. Methods of blood flow Modelling. Mathematical Modelling of Natural Phenomena. 2016;**11**(1):1-25

[33] Jahandardoost M, Fradet G, Mohammadi H. A novel computational model on the hemodynamics of the bileaflet mechanical heart valves. Proceedings of the Institution of Mechanical Engineers, Part H. 2015;**229**(3):232-244

[34] Raustin R, Mohammadi H. Towards multiscale modeling of wave propagation in arteries. Journal of Mechanics in Medicine and Biology. 2015;**16**(2):1650027(1-10)

[35] Jahandardoost M, Fradet G, Mohammadi H. Hemodynamic study of the elliptic St. Jude medical valve; a computational study. Proceedings of the Institution of Mechanical Engineers, Part H. 2016;**230**(2):85-96

[36] Crampton L. How Blood Clots: Platelets and Coagulation Cascade. 2017. Retrieved from: https://owlcation.com/stem/How-Does-Blood-Clot-and-What-Causes-Coagulation

[37] European Haemophilia Consortium: Platelet Disorders. Retrieved from: https://www.ehc.eu/bleeding-disorders/patelet-disorders/

[38] Mohammadi H, Nestor B, Fradet G. Simulation of anastomosis in coronary artery bypass surgery. Cardiovascular Engineering and Technology. 2016;7:432-438

[39] Sharifikia D, Yafia MS, Fradet G, Mohammadi H. The design and fabrication of a prosthetic aortic root made of hydrogel biomaterials. Journal of Medical and Biological Engineering. 2017. DOI: 10.1007/s40846-017-0290-9

[40] Jahandardoost M, Ohlmann L, Fradet G, Mohammadi H. Effect of the heart rate on the hemodynamics of the elliptic St. Jude medical valve; a computational study. Journal of Mechanics in Medicine and Biology. 2018;**18**(2):1850014

[41] Mohammadi H. A numerical method to enhance the accuracy of the mass – spring system for soft tissue deformation, Journal of Applied Biomechanics. 2009;**25**(3):271-278

[42] Mohammadi H, Ahmadian MT, Wan WK. Time-dependent analysis of leaflets in mechanical aortic bileaflet valve in closing phase using the finite strip method. Journal of Medical Engineering & Physics. 2006;**28**(2):122-133

[43] Mohammadi H, Mequanint K. Effect of stress intensity factor in evaluation of instability of atherosclerotic plaque. Journal of Mechanics in Medicine and Biology. 2014;**14**(5):1450072(1-15)

[44] Fung YC. Mechanical properties of blood vessels. In: Biomechanics. New York, NY: Springer; 1981. pp. 261-301

[45] T. R. Canfield and P. B. Dobrin, "Mechanics of Blood Vessels."

[46] Mohammadi H, Mequanint K, Herzog W. Multiscale modeling of O_2 transport in articular cartilage. Biophysical Journal. 2012;**102**(3):592a

[47] Bahramian F, Mohammadi H. Combined experimental and numerical studies of heat transfer in a non-equipped transporter for hypothermic conditions. Journal of Mechanics in Medicine and Biology. 2012;**12**(3):1250057 (1-11)

[48] Mohammadi H, Mequanint K. An inverse numerical approach for modeling aortic heart valve leaflet tissue oxygenation. Cardiovascular Engineering and Technology. 2012;**3**(1):73-79

[49] Charalambous HP, Roussis PC, Giannakopoulos AE. Viscoelastic dynamic arterial response. Computers in Biology and Medicine. 2017;**89**(Supplement C):337-354

[50] Corbett SC, Ajdari A, Coskun AU, Nayeb-Hashemi H. Effect of pulsatile blood flow on thrombosis potential with a step wall transition. ASAIO Journal. 2010;**56**(4):290-295

[51] Hashimoto S, Manabe S, Matsumoto Y, Ikegami K, Tsuji H. The effect of pulsatile shear flow on thrombus formation and hemolysis. In: 22nd Annual EMBS International Conference; Chicago. 2000. pp. 2461-2462

[52] Fallon AM, Dasi LP, Marzec UM, Hanson SR, Yoganathan AP. Procoagulant properties of flow fields in stenotic and expansive orifices. Annals of Biomedical Engineering. 2008;**36**(1):1-13

[53] Zakaria MS, Ismail F, Tamagawa M, Aziz AFA, Wiriadidjaja S, Basri AA, et al. Review of numerical methods for simulation of mechanical heart valves and the potential for blood clotting. Medical & Biological Engineering & Computing. 2017;**55**(9):1519-1548

[54] Yoganathan AP. Cardiac valve prostheses. In: Bronzino EJD, editor. The Biomedical Engineering Handbook, Chapter 127. 2nd ed. Boca Raton: CRC Press; 2000

[55] Krishnan S, Udaykumar HS, Marshall JS, Chandran KB. Two-dimensional dynamic simulation of platelet activation during mechanical heart valve closure. Annals of Biomedical Engineering. 2006;**34**(10):1519-1534

[56] Bluestein D, Rambod E, Gharib M. Vortex shedding as a mechanism for free emboli formation in mechanical heart valves. Journal of Biomechanical Engineering. 2000;**122**:125-134

[57] Yun BM, Aidun CK, Yoganathan AP. Blood damage through a bileaflet mechanical heart valve: A quantitative computational study using a multiscale suspension flow solver. Journal of Biomechanical Engineering. 2014;**136**(10):17

[58] Hong T, Kim CN. A numerical analysis of the blood flow around the bileaflet mechanical heart valves with different rotational implantation angles. Journal of Hydrodynamics, Series B. 2011;**23**(5):607-614

[59] Grigioni M, Morbiducci U, Avenio GD, Di G, Costantino B, Gaudio D. A novel formulation for blood trauma prediction by a modified power-law mathematical model. Biomechanics and Modeling in Mechanobiology. 2005;**4**(4):249-260

[60] Xenos M, Girdhar G, Alemu Y, Jesty J, Slepian M, Einav S, et al. Device thrombogenicity emulator (DTE)—Design optimization methodology for cardiovascular devices: A study in two bileaflet MHV designs. Journal of Biomechanics. 2010;**43**(12):2400-2409

[61] Sheriff J, Soared JOS, Xenos M, Jesty J, Bluestein D. Evaluation of shear-induced platelet activation models under constant and dynamic shear stress loading conditions relevant to devices. Annals of Biomedical Engineering. 2013;**41**(6):1279-1296

[62] Simon HA, Ge L, Sotiropoulos F, Yoganathan AP. Numerical investigation of the performance of three hinge designs of bileaflet mechanical heart valves. Annals of Biomedical Engineering. 2010;**38**(11):3295-3310

[63] Biasetti J, Hussain F, Gasser TC. Blood flow and coherent vortices in the normal and aneurysmatic aortas: A fluid dynamical approach to intra-luminal thrombus formation. Journal of the Royal Society Interface. 2011;**8**:1449-1461

[64] Naimah W, Ab W, Balan P, Sun Z, Miin Y. Prediction of thrombus formation using vortical structures presentation in Stanford type B aortic dissection: A preliminary study using CFD approach. Applied Mathematical Modelling. 2016;**40**(4):3115-3127

[65] Tamagawa M, Kaneda H, Hiramoto M, Nagahama S. Simulation of thrombus formation in shear flows using lattice Boltzmann method. Artificial Organs. 2009;**33**(8):604-610

Section 3

Analytical Biomechanics

Chapter 3

An Introduction to Survival Analytics, Types, and Its Applications

Sheik Abdullah Abbas, Selvakumar Subramanian, Parkavi Ravi, Suganya Ramamoorthy and Venkatesh Munikrishnan

Abstract

In today's world, data analytics has become the integral part of every domain such as IOT, security, healthcare, parallel systems, and so on. The importance of data analytics lies at the neck of what type of analytics to be applied for which integral part of the data. Depending upon the nature and type of data, the utilization of the analytical types may also vary. The most important type of analytics which has been predominantly used up in health-care sector is survival analytics. The term survival analytics has originated from a medical domain of context which in turn determines and estimates the survival rate of patients. Among all the types of data analytics, survival analytics is the one which entirely depends upon the time and occurrence of the event. This chapter deals with the need for survival data analytics with an explanatory part concerning the tools and techniques that focus toward survival analytics. Also the impact of survival analytics with the real world problem has been depicted as a case study.

Keywords: classification, data analytics, statistics, survival analytics, prediction, parametric models

1. Introduction to survival analytics

Survival analysis refers to a branch of statistical analysis domain that evaluates the effect of predictors on *time until an event*, rather than the *probability of an event*, occurs. It is used to analyze data in which the time until the event is of interest. As the name indicates, this method has origins in the field of medical research for evaluating the impact of medicines or medical treatment on time until death. Survival analysis is also known as reliability analysis in the engineering discipline, duration analysis in the economics discipline, and event history analysis in the sociology discipline.

The term is originated from a medical context in which it has been used to estimate the survival rate of patients. Data classification can be dealt explicitly with the process and paradigms available in survival analytical models [1]. The process of survival analytics can be explored through various techniques such as:

- Life tables
- Kaplan-Meier analysis
- Survivor and hazard function rates
- Cox proportional hazards regression analysis
- Parametric survival analytic models
- Survival trees
- Survival random forest

2. Metrics for measurement in developing survival models

The process of survival analytics mainly depends on time and occurrence of the event. In survival analytics, time-varying covariates are the variables considered, which change with accordance to the occurrence of the event [2]. The process of survival analytics can be signified and measured using the following measurements:

1. Event time distribution

The event time distribution corresponding for an event to occur with respect to the time function t is defined in Eq. (1) as

$$f(t) = \lim_{\Delta t \to 0} \frac{p(t \le T < t + \Delta t)}{\Delta t} \tag{1}$$

where t is the occurrence of an event at a time t and Δt is the small change in time t with accordance to the event.

The distribution varies with accordance to a small change in time for the event that tends to happen for the given function $f(t)$.

2. Cumulative event time distribution

The cumulative event time distribution for the given function $f(t)$ is defined in Eq. (2) as

$$F(t) = P(T \le t) = \int_0^t f(u)du \tag{2}$$

The time period is estimated to be from 0 to t.

3. Survival function

The survival function provides the probability estimate in which the corresponding object of life will have existence beyond the period of time t. This measure is also termed to be the survivor function or reliability function. The measure can be estimated using the following Eq. (3):

$$S(t) = 1 - F(t) = P(T > t) = \int_0^\infty f(u)du \tag{3}$$

For the condition $S(0) = 1$ and $S(\infty) = 0$, the following relationship holds:

$$f(t) = \frac{dS(t)}{dt} \tag{4}$$

4. Hazard function

The hazard function is also termed to be the hazard rate or the value of mortality, which is then the ratio among the probability density function and the survivor function which is depicted in Eq. (5) as

$$h(t) = \frac{f(t)}{S(t)} \tag{5}$$

where $h(t)$ is the hazard function, $f(t)$ is the probability density function, and $S(t)$ is the survival function.

3. Model classification in survival analytics

The process behind survival analytics is different when compared to predictive and descriptive analytics [3]. Here, the time component is an important factor which efficiently determines the success or failure of a model. The following **Figure 1** illustrates the model to be classified under survival analytics [4]. Different sorts of functions are adaptable with different models based on the metric to be used with time as a component for the event to occur. The main target is to determine the right model to be chosen for the observed survival analytic data. In parametric model analysis, the survival curve depends only on the shape of the model with its function value.

The shape of the model can be estimated with regard to the characteristics of a nonparametric model. As an outcome, the shape of the hazard function also varies with regard to time. Some of the examples corresponding to hazard shapes are:

- Increasing hazard
- Decreasing hazard
- Constant hazard
- Convex bathtub-shaped hazard

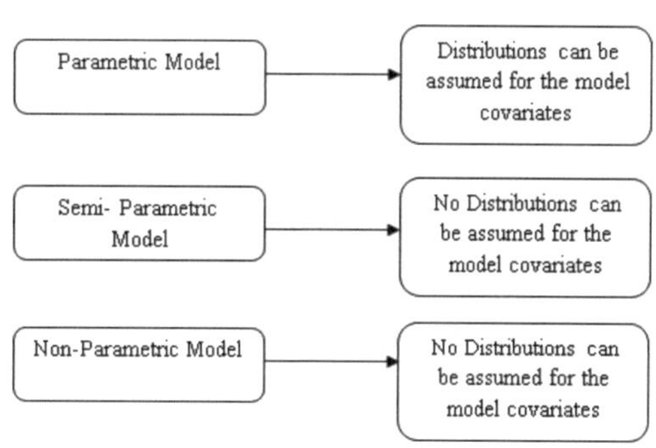

Figure 1.
Model classification in survival analytics.

Parametric survival analysis can be understood with the available forms of distributions. The distributions usually convey the efficacy in terms of probability curve with time analysis. The following are the distributions used to estimate the output measured by the survival curve:

1. Normal distribution

2. Weibull distribution

3. Exponential distribution

4. Lognormal distribution

5. Gamma distribution

6. Uniform distribution

7. Log-logistic distribution

The following illustrations provide an overview with regard to each of the distributions in detail:

4. Parametric survival analytical models

4.1 Exponential distribution

The exponential distribution is also known as negative exponential distribution. Exponential distribution is defined as a process in which events occur continuously and independently at a constant average rate. The exponential distribution is defined as

$$f(t) = \lambda e^{-\lambda t} \tag{6}$$

The survivor function is then estimated as

$$S(t) = \lambda e^{-\lambda t} \tag{7}$$

The hazard rate is then estimated as

$$h(t) = \frac{f(t)}{S(t)} \tag{8}$$

$$h(t) = \lambda \tag{9}$$

Hence, from Eq. (9), it should be noted that the hazard rate is independent of time and therefore the risk corresponding to the event remains to be same. The following **Figures 2** and **3** illustrate the event time with respect to the hazard rate.

4.2 Weibull distribution

The Weibull distribution is defined as a continuous probability distribution. The expression is defined in Eq. (10) as

$$f(t) = k\rho\,(\rho t)^{k-1}\exp\left[-(\rho t)^{k}\right] \tag{10}$$

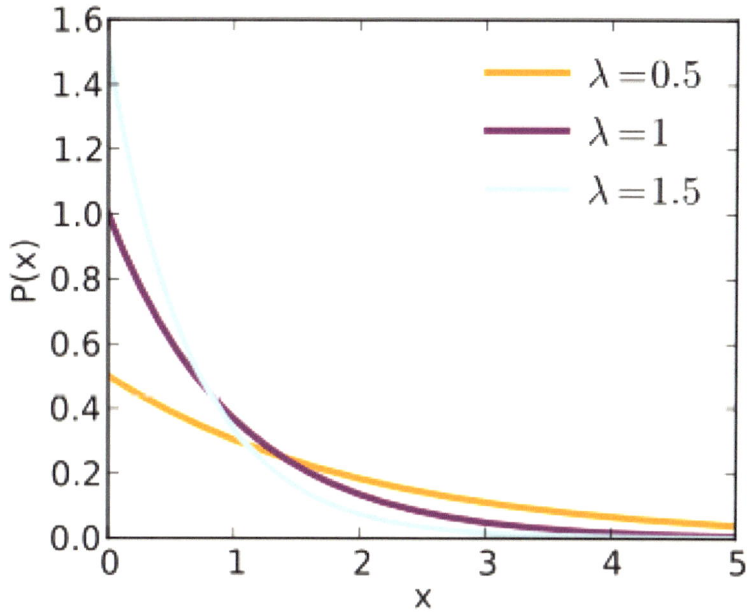

Figure 2.
Probability density function.

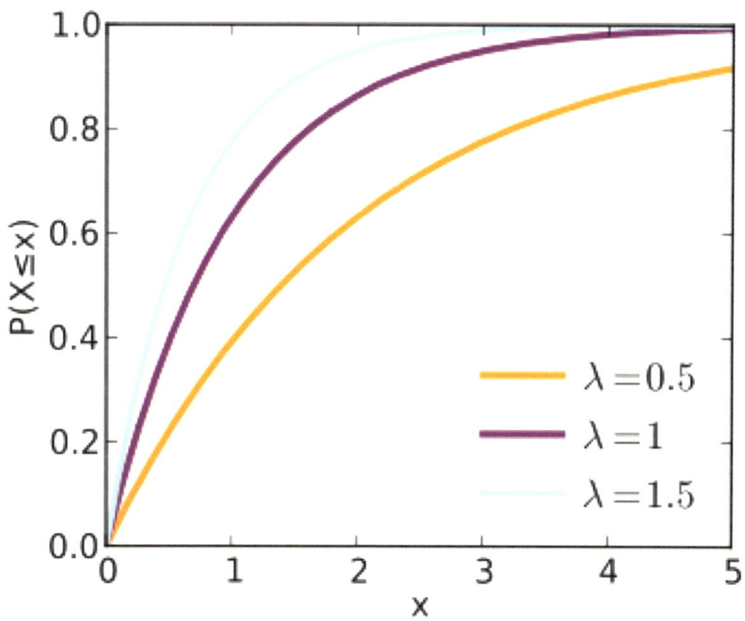

Figure 3.
Cumulative distribution function.

The evaluation of hazard rate is given as

$$h(t) = k\rho\,(\rho t)^{k-1} \tag{11}$$

Hence, for this case, the hazard rate depends on time which can be in either increasing or decreasing mode. The following **Figures 4** and **5** depict the value of hazard rate with respect to time t.

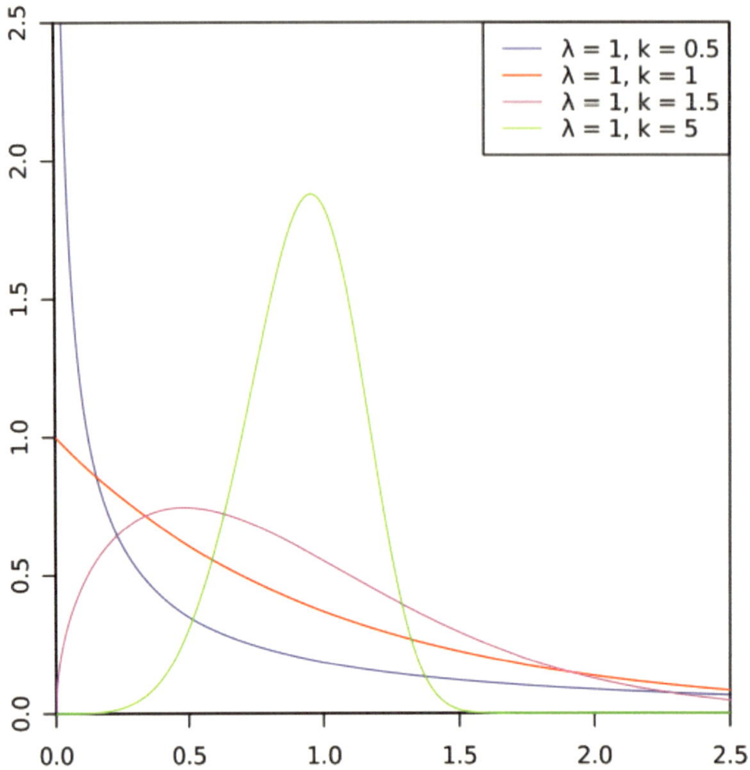

Figure 4.
Probability distribution function.

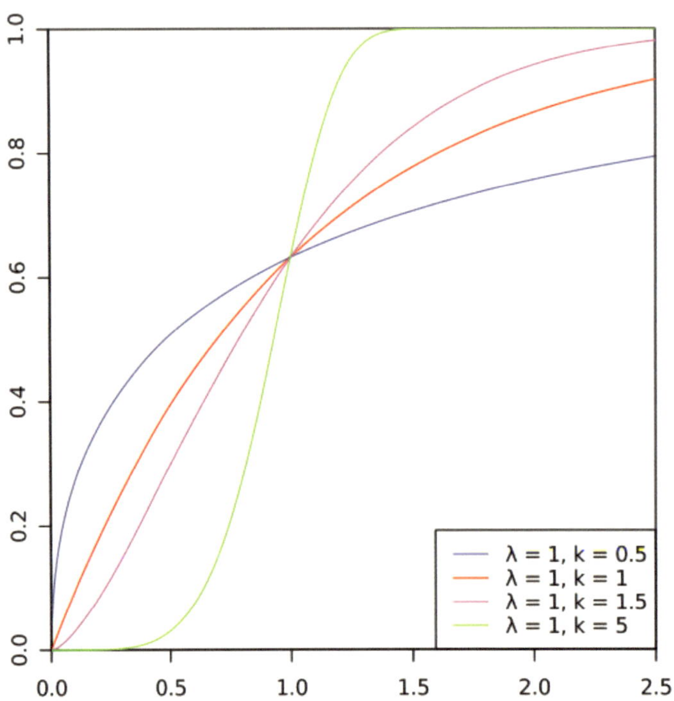

Figure 5.
Cumulative distribution function.

4.3 Log-logistic distribution

The log-logistic distribution is defined as a continuous probability distribution with negative random variable. In economics discipline log-logistic distribution is also known as the Fisk distribution in economics. Log-logistic is a continuous probability distribution for a nonnegative random variable. The following **Figures 6** and 7 depict the distribution of log-logistic model.

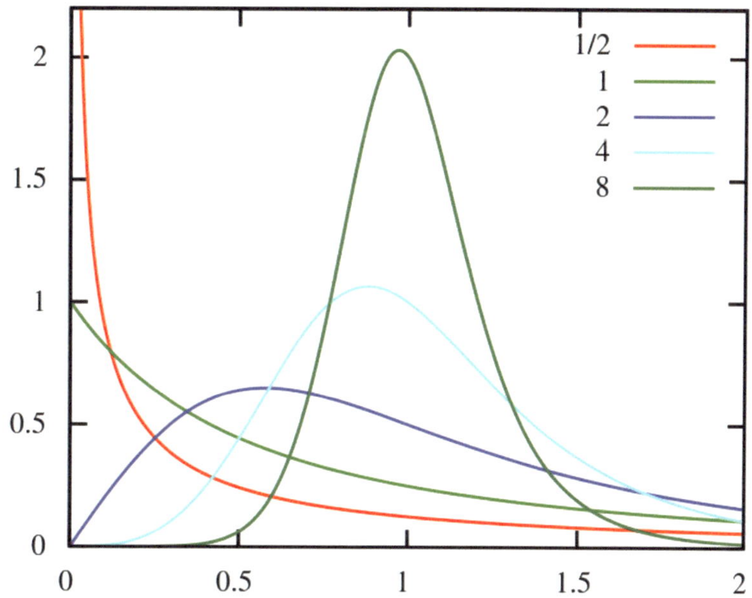

Figure 6.
Probability density function.

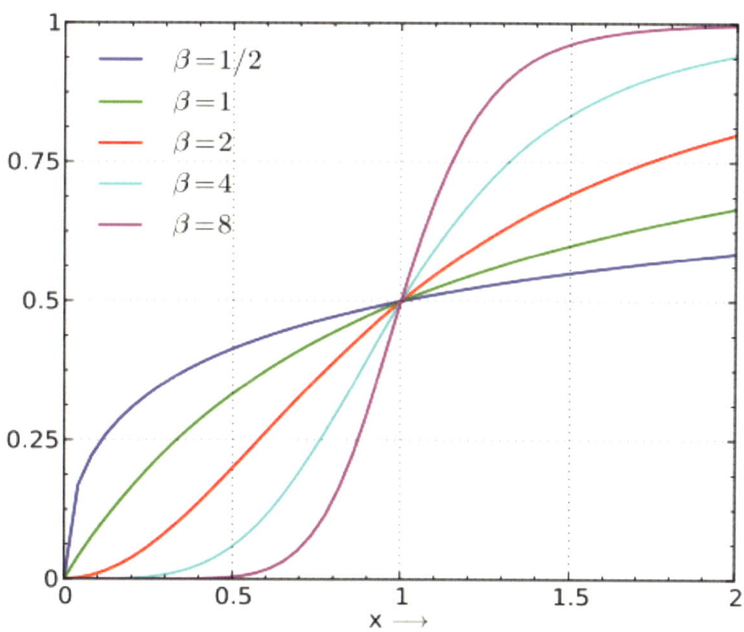

Figure 7.
Cumulative distribution function.

5. Kaplan-Meier analytics and Cox regression model

The Kaplan-Meier test is broadly used within the pharmaceutical industry for specifying the expiry data for clinical drug in health-care sector, monitoring the effects of drugs and their gestures on recovery time or critical time. The Kaplan-Meier test is a statistical method that really works well for effective cancer treatments. This test determines the patient's survival time between two groups. For clinical drug examination, a successful test indicates that the group of people taking the new drug has a shorter time to improvement or death than the group of people taking a place [5].

If the value of censoring is not available, then the value of the KM estimator for $S(t)$ is found to be the same proportional value with respect to t. If the value of censoring is present, then the following steps are followed:

Step 1: Order the event times in ascending order of levels $t_1 < t_2 < t_3 < \cdots < t_k$.

Step 2: At each time t_j, there are about n_j individuals who are subjected to risk of the event.

Step 3: Let d_j be the number of individuals who die at t_j (churn, respond).

Therefore, the KM estimator is defined as

$$S(t) = \prod_{J:t_1 \le t} \left(1 - \frac{d_j}{n_j}\right) = S(t-1) \bullet \left(1 - \frac{d_t}{n_t}\right) \tag{12}$$

If there exists uniqueness in event times, then the KM estimator is measured using the life table for grouping of event times as expressed as

$$S(t) = \prod_{J:t_1 \le t} \left(1 - \frac{d_j}{n_j - c_j/2}\right) \tag{13}$$

where

n_j is the number of individuals at risk.

d_j is individuals who die at a specified time.

The implementation of KM estimator can be extended by applying statistical tests such as hypothesis testing, Wilcoxon test, and likelihood ratio test. With exploratory data analytics along with KM estimator, the patterns and insights can be determined more efficiently from the data.

The second popular survival analysis method used for prediction is Cox. It is also known as the Cox model but often referred to as Cox regression. It is more popular in the Web of Science. More than 38,000 articles are cited indexing the Cox regression method.

There are some other statistical/analytical methods available that can predict time until an event, but survival analysis methods have the unique feature of considering the past history/experiences. Although these latter cases do not have a date for the target event, they are an integral part of the analysis. The terminology used in survival analysis is called censored cases.

Another formal definition for survival analysis is, it is basically defined as a set of methods for analyzing data where the outcome variable is the time/instance until the occurrence of an event of interest. The event can be an uncertainty accident, death, occurrence of a disease, or planned ones—marriage, divorce, etc. The time to event or survival time can be measured in various scales of time periods (days, weeks, years, etc.).

For example, if the event of interest is mild heart attack, then the survival time can be the time in years until a person develops a heart attack. Choose any survival methods that are discussed above. In survival analysis, time is a primary factor.

The advantage of Cox regression over Kaplan-Meier is that it can accommodate any number of predictors, i.e., chances of getting heart attack, rather than group membership only. As is the case for all regression methods, there are two potential benefits of analysis using Cox regression: *predictor ranking*, with each predictor's effect measured greater than the predicator's threshold effect or less than the predicator's threshold effect and the ability to *make predictions* with the regression results. Predictor rankings facilitate the analyst to recognize the factors that have the most influence on time to an event, and the regression results can be used to estimate the amount of until an event for a specific profile of any subject [6].

5.1 Different types of censoring

Data can be either right, left, or interval censored. It is the sum of defined time t_0, and the event of interest takes place at $t_0 + t$, where t is an unknown factor and the event is only known to have occurred at $t_0 + c$ and the data is censored with a censored time, c.

Right censoring is the most common, occurring when the true event time is greater than the censored time, when $c < t$. It often arises when the event of interest has not occurred by the end of study and the subject has been lost to follow-up.

Left censoring is the opposite, occurring when the true event time is less than the censored time, when $c > t$.

Interval censoring is a concatenation of the left and right censoring, when the time is known to have occurred between two time points: $c_1 < t < c_2$.

Censoring is an important matter in survival analysis, signifying a particular type of missing data. Censoring is a random and non-informative study, and it is usually required in order to avoid bias in a survival analysis The interpretation of Cox regression and Kaplan results depends two factors: positive (e.g., a sale) or negative (e.g., product failure).

6. Case study for churn prediction

The following graphical illustrations depict the implementation of churn prediction and model deployment using RapidMiner. The algorithm used for analysis is decision tree [7]. The implementation has been done with the lift chart analysis with evaluation in performance metrics. The attributes in the dataset includes person ID, churn status, gender, age, region code, transaction count, average balance, and total accounts.

The case study majorly explores with an application that is most probably used up with churn analysis. Nowadays, churn prediction is majorly analyzed in most of the industries to track the historical learning with the customers. The entire customer demographic data is analyzed day to day with regard to the maintenance of business relationships, customer transactions, products purchased, and the survey that has been obtained with regards to the business attractions. To make an exploration in this application, we have used up RapidMiner tool for the entire survival rate estimation and analysis of customers in an organization. The above **Figure 8** provides the selection of application with regard to churn prediction for estimating the survival rate of customers.

RapidMiner is one of the good statistical and analytical tools which is mostly practiced in industries and academic institutions. Rapid miner provides a good insight for statisticians and mathematical experts to observe the insights and patterns that lie within the given data. The following **Figure 9** explores the analytical results observed with RapidMiner.

All the incorporations in RapidMiner are made through the process connection through wires. The workflow of each process is written through Java. The process diagram depicts the step-by-step flow of algorithmic model development through drag option. **Figure 10** provides a complete overview with regard to the process creation for churn prediction analysis. The algorithm used for the development of the model is decision tree classification algorithm [8]. Decision tree algorithm provides a tree-like structure in a top-down fashion with a single root node and a number of

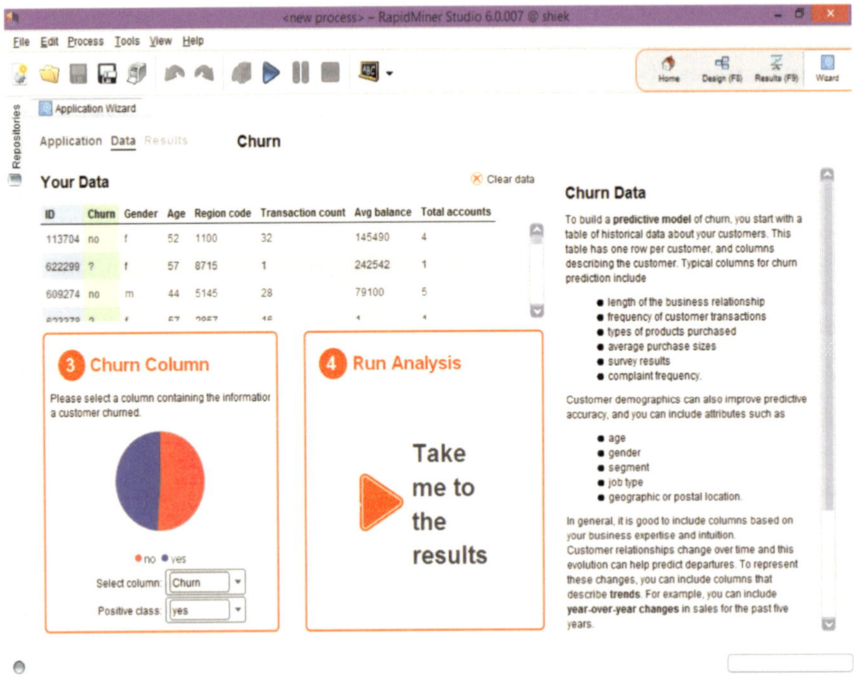

Figure 8.
Data selection for churn analysis.

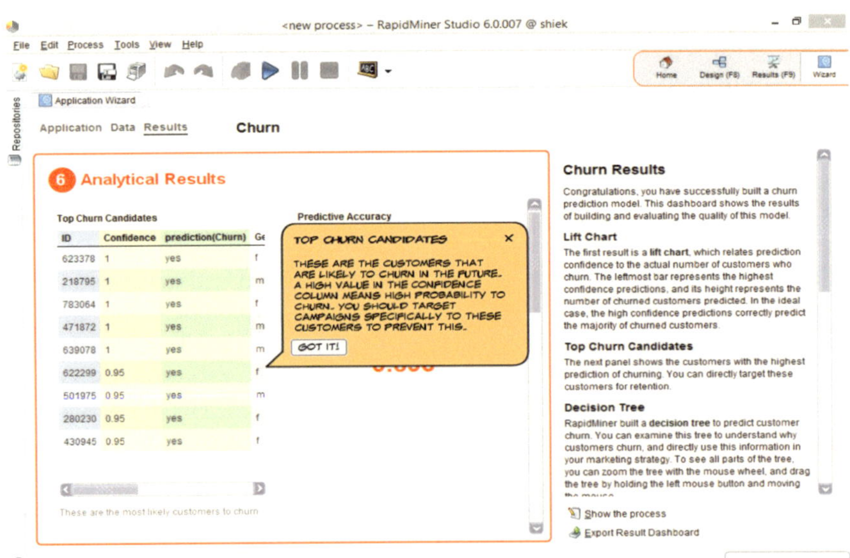

Figure 9.
Analytical results observed for churn model.

leaf nodes with a terminating condition. The working of the algorithm depends on the splitting criterion to be used up for analysis (**Figures 11** and **12**).

Lift chart shows the effectiveness of the predictive model in which it has to be developed. It generally provides the ratio between the predicted values to that of the actual one. In **Figure 13** for churn analysis, the chart provides the ratio between the confidence value and the count observed for churn analysis. Thus, churn prediction is employed for tracking the survival rate of customers with survival analytics. Survival analytics model can be deployed more efficiently for tracking the rate of patients in medical domain. The realm of health informatics lies at the heart of existence of subjects concerned with specific disease. The existence and

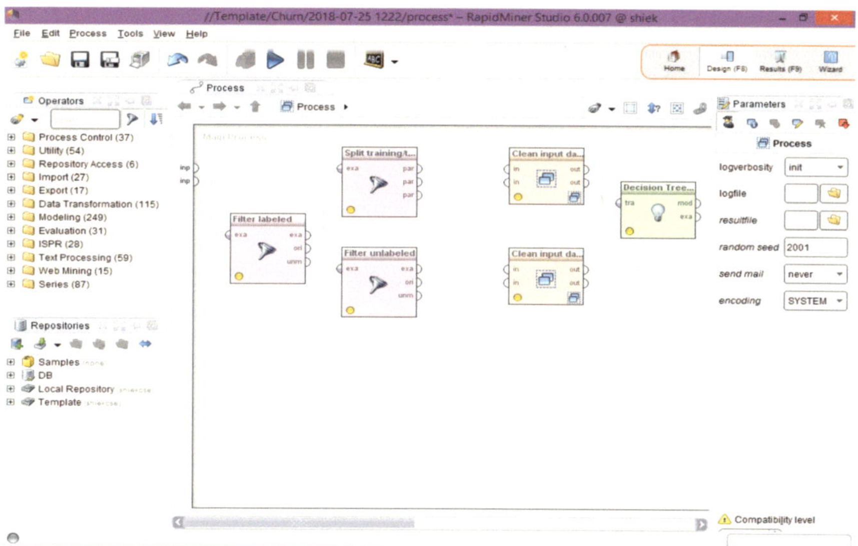

Figure 10.
Process diagram in a design view.

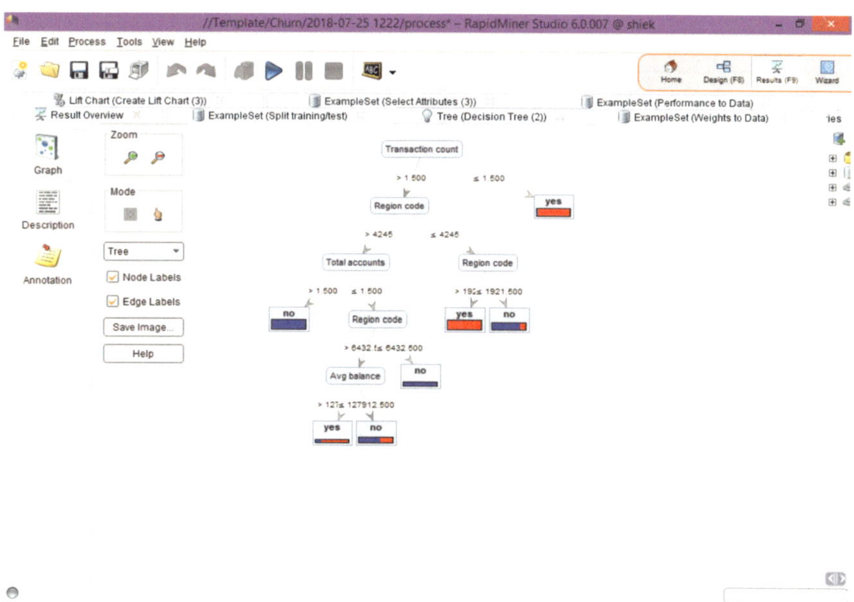

Figure 11.
Generated tree with decision tree algorithm.

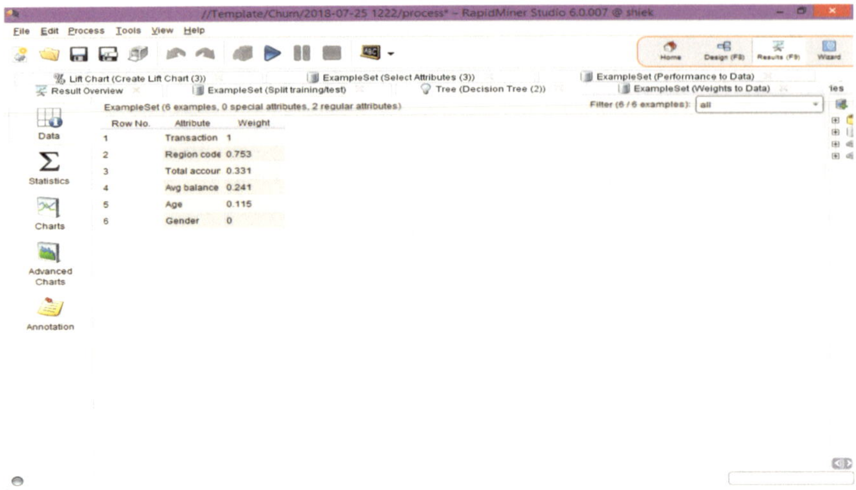

Figure 12.
Performance evaluation.

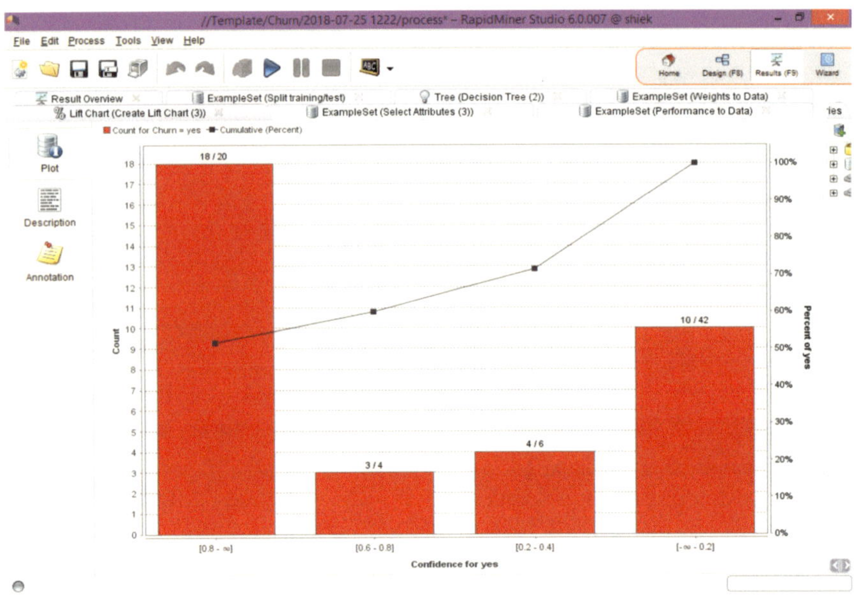

Figure 13.
Lift chart for churn analysis.

the nonavailability of subjects with regard to the specific disease can be learnt with patterns and explorations through survival analytics models.

7. Case study using Kaplan-Meier analytics

Consider there are about 200 subjects of patient's records which have been tracked over a period of time. The tracking is made in such a way that the total number of patients has to be confirmed with first year, second year, and third year, and so on. If all the subjects have existed for the given duration, then there will not be a case for probability of occurrence with regard to each of the subjects. To illustrate this complicated situation, consider the following scenario in **Table 1**.

Condition:

1. Out of 200 subjects, 6 became unavailable and 10 have been found to be dead at the end of the first year.

2. With the remaining subjects, 6 became unavailable and 20 have been found to be dead at the end of the second year.

3. With the remaining subjects, 6 became unavailable and 30 have been found to be dead at the end of third year.

4. With the remaining subjects, 6 became unavailable and 40 have been found to be dead at the end of fourth year.

5. With the remaining subjects, 6 became unavailable and 50 have been found to be dead at the end of fifth year.

Time period	At risk	Become unavailable (censored)	Died	Survived
Year 1	200	6	10	?
Year 2	?	6	20	?
Year 3	?	6	30	?
Year 4	?	6	40	?
Year 5	?	6	50	?

Table 1.
Scenario of individuals for a specified disease.

Time period	At risk	Become unavailable (censored)	Died	Survived
Year 1	200	6	10	190
Year 2	184	6	20	164
Year 3	158	6	30	128
Year 4	122	6	40	82
Year 5	76	6	50	26

Table 2.
Observed solution as per Kaplan-Meier condition.

Time period	At risk	Become unavailable (censored)	Died	Survived	Kaplan-Meier survival probability estimate
Year 1	200	6	10	190	(190/200) = 0.95
Year 2	184	6	20	164	(190/200) × (164/184) = 0.84
Year 3	158	6	30	128	(190/200) × (164/184) × (128/158) = 0.70
Year 4	122	6	40	82	(190/200) × (164/184) × (128/158) × (82/122) = 0.46
Year 5	76	6	50	26	(190/200) × (164/184) × (128/158) × (82/122) × (26/76) = 0.15

Table 3.
Kaplan-Meier probability estimate.

For this scenario we can determine the list of individuals who are all became unavailable at the end of the given time period. Use Kaplan-Meier analytics to determine the individuals who are at risk and what would be the probability estimate for the individuals survived at the end of the fifth year.

Solution:

Step 1: Kaplan-Meier suggested that the subjects that became unavailable during the given time period can be counted among with those who survive through the end but are removed or deleted from the total number of individuals who are subjected to risk for the next given time period. With these conventions, the formulation is described in **Table 2**.

Hence, from **Table 2**, it has been observed that at the end of fifth year, 26 individuals have survived from the set of 200 individuals who were subjected to a specified disease. The next is to determine the Kaplan-Meier probability estimate for each of the time intervals t with regard to the conditional probability. The following **Table 3** provides the probability estimate for 5 years of risk analysis.

From **Table 3**, it has been observed that at the end of fifth year, the conditional probability estimate was found to be 0.15% of individuals. Hence, from the perspective of survival probabilistic estimate, we can determine the existence rate of individuals for the given time period t.

Author details

Sheik Abdullah Abbas[1*], Selvakumar Subramanian[2], Parkavi Ravi[1], Suganya Ramamoorthy[1] and Venkatesh Munikrishnan[3]

1 Department of Information Technology, Thiagarajar College of Engineering, Madurai, Tamil Nadu, India

2 Department of Computer Science and Engineering, GKM College of Engineering and Technology, Chennai, Tamil Nadu, India

3 Department of General Medicine, Theni Government Medical College and Hospital, Theni, Tamil Nadu, India

*Address all correspondence to: asait@tce.edu

References

[1] Sheik Abdullah A, Gayathri N, Selvakumar S, Rakesh Kumar S. Identification of the Risk Factors of Type II Diabetic Data Based Support Vector Machine Classifiers upon Varied Kernel Functions. Lecture Notes in Computational Vision and Biomechanics [Internet]. Switzerland: Springer International Publishing; 2018. pp. 496-505. Available from: http://dx.doi.org/10.1007/978-3-319-71767-8_42

[2] Tsujitani M, Baesens B. Survival analysis for personal loan data using generalized additive models. Behaviormetrika [Internet]. Behaviormetric Society of Japan; 2012; **39**(1):9-23. Available from: http://dx.doi.org/10.2333/bhmk.39.9

[3] Sheik Abdullah A, Selvakumar S, Abirami AM. An Introduction to Data Analytics. Handbook of Research on Advanced Data Mining Techniques and Applications for Business Intelligence [Internet]. Hershey, USA: IGI Global; 1-14. Available from: http://dx.doi.org/10.4018/978-1-5225-2031-3.ch001

[4] Cox DR. Regression Models and Life-Tables. Breakthroughs in Statistics [Internet]. Switzerland, New York: Springer; 1992;527-541. Available from: http://dx.doi.org/10.1007/978-1-4612-4380-9_37

[5] Banasik J, Crook JN, Thomas LC. Not if but When will borrowers default. The Journal of the Operational Research Society [Internet]. JSTOR; 1999 Dec;**50**(12):1185. Available from: http://dx.doi.org/10.2307/3010627

[6] Crowder M. Classical Competing Risks. UK: CRC Press, Taylor & Francis Group, an Informa Group company; 2001 May 11. Available from: http://dx.doi.org/10.1201/9781420035902

[7] Abdullah AS, Selvakumar S, Karthikeyan P, Venkatesh M, et al. Comparing the efficacy of decision tree and its variants using medical data. Indian Journal of Science and Technology [Internet]. Indian Society for Education and Environment. 2017 May 1; **10**(18):1-8. Available from: http://dx.doi.org/10.17485/ijst/2017/v10i18/111768

[8] Sheik Abdullah A, Suganya R, Selvakumar S, Rajaram S. Data Classification. Handbook of Research on Advanced Data Mining Techniques and Applications for Business Intelligence [Internet]. Hershey, USA: IGI Global; 34-51. Available from: http://dx.doi.org/10.4018/978-1-5225-2031-3.ch003

Section 4

Biomechanics of Human Eyes

Chapter 4

Biomechanics of Eye Globe and Methods of Its Study

Irina Bubnova

Abstract

Knowledge of biomechanical properties of eye globe is necessary both for correct selection of candidates for refractive surgery and right choice of operative intervention parameters. No less important, it is for corneal ectatic disease diagnostics and monitoring. Also it gives inestimable contribution for interpretation of intraocular pressure (IOP) indices especially in cases with irregular eye shape or after past corneal surgical procedures. Moreover, it allows studying injury mechanism by glaucoma process on optic nerve head fibers. Above it, scleral biomechanical properties research is necessary for the investigation of pathophysiologic factors of myopia manifestation and progression. This chapter is devoted to review of existed to date methods of study of eye fibrous tunic biomechanical properties. It describes mathematical, experimental, and clinical models provided evaluation of unsearchable by direct measurement parameters. It also observes effective technics of impact on both sclera and cornea with the aim of correction of its biomechanical condition.

Keywords: corneal biomechanics, refractive surgery, LASIK, keratokonus, IOP

1. Introduction

The cornea and the sclera are two conjugated quasi-spherical segments with unequal curvature radii; together they form corneoscleral (fibrous) tunic—the supporting structure of the eye capsule. Their mechanical properties play a crucial role of holding together the inner ocular structures. Despite them both being composed of connective tissue, they differ in physical (particularly, optical) and biomechanical properties [1].

The cornea is the anterior part of the fibrous tunic of the eyeball, and it takes up 1/6 of its length. Despite it being relatively thin, its main function is protection—assured by its high durability. But the cornea also participates in light ray refraction, making up an important part of the visual apparatus; as such, it is characterized by high optical homogeneity and complete transparence.

The cornea is an anisotropic, inhomogeneous structure; it mainly consists of highly specific connective tissue formed by parallel collagen fibrils that extend from one limb to another and act as load supporting elements [2].

The sclera takes up the other 5/6 of the eye length and represents the posterior part of the fibrous tunic of the eyeball. Scleral tunic is the main supporting structure of the eyeball; it consists of dense collagen fibers. In contrast to cornea, the sclera has high dispersive power due to its chaotically distributed fibrils and fibers, which prevents light from entering the eye cavity from the side. In natural conditions, in the living eye, scleral elements are constantly in a strain-stress state determined by intraocular pressure and mechanical properties of the scleral tissue, as well as by anisotropy and inhomogeneity of these properties [3].

Studying the biomechanical properties of the cornea is relevant for certain clinical needs associated with the appearance of new biomechanics examination methods, as well as the need to diagnose and monitor ectatic diseases of the cornea, to adequately select the parameters for keratorefractive surgeries, to correctly interpret the intraocular pressure (IOP) values, and, consequently, to appropriately assess IOP and monitor glaucomatous process.

In addition, conducting studies on the biomechanical properties of the sclera is a necessary step in the research of pathogenic factors relevant for occurrence and progression of myopia, as well as finding effective means and methods of influencing the sclera in order to correct its biomechanical state.

However, the lack of standardized terminology and uniform classification hurts the ability to compare research results and consequently hinders their introduction into the knowledge area of ophthalmology.

2. Classification of approaches to study the biomechanics of the eye

In accordance with different approaches, eye biomechanics can be divided into the following types:

1. theoretical;

2. physical (i.e., experimental); and

3. clinical.

2.1 Theoretical biomechanics of the eye

Theoretical biomechanics is a science that employs mathematical methodology and mathematical analysis. As applied to ophthalmology, it handles with specific physical constants characterizing elasticity, strength, and other mechanical parameters of the tissues (usually measured *in vitro*).

The main theoretical approach is mathematical modeling. The research may target separate structures of the eyeball and the tunic, or the eye in its entirety. It can also include modeling of physiological or pathological processes, changes induced by specific stimuli or effects of surgical treatment.

The results obtained from modeling can be used in experimental and clinical studies. In turn, all models are based on the figures acquired in experiments or from clinical diagnostic.

Disadvantages of the theoretical approach in studying eye biomechanics are associated with structural complexity of the eyeball, inhomogeneity, and variability of morphology of the ocular structures, and dependence on the technological advancement of experimental and clinical research methods.

2.2 Physical (experimental) biomechanics of the eye

Experimental biomechanics of the eye is based on studying individual tissues and the eyeball as a whole *in vitro* or by conducting animal experiments using physical methods. It is the most developed subdiscipline of biomechanics with many years of research history. Capabilities of the approach are limited by post-mortem changes in eye tissues, and anatomical and physiological differences between humans and animals. The main purpose of experimental studies is to find potentially useful methods of studying biomechanical properties of eye tissues in clinical environment and to acquire data for mathematical modeling. The main advantages of experimental researches are the absence of restrictions for employed methods and approaches, and the choice of which is only limited by technological and scientific advancement.

Methods of experimental biomechanics allow measurement of a big number of physical parameters of the cornea:

- Young's modulus (E),
- Poisson's ratio (μ),
- Durability (σ),
- Deformation capacity (Σ), etc.

However, they do not fully reflect the properties of fibrous tunic of the eye *in vivo*.

2.3 Clinical biomechanics of the eye

Clinical biomechanics of the eye studies the influence of biomechanical properties of the fibrous tunic on the results of diagnostics, development, and treatment of various eye diseases. Clinical biomechanics operates on data obtained with specialized examination methods used in ophthalmology (*in vivo*) that characterize biomechanical properties of the fibrous tunic. Its research subject is strictly the eyeball as a whole, only allowing arbitrary delineation of the internal structures. This complicates the interpretation of data. However, in order to improve diagnostics and treatment of eye diseases, clinical methods for eye biomechanics need to have higher priority in research and development.

The following corneal parameters can be measured clinically:

- Friedenwald's rigidity coefficient;
- Corneal hysteresis;
- Corneal resistance factor;
- Coefficient of elasticity;
- Corneal deformation.

The biggest number of already existed studies are dedicated to investigation of biomechanical properties of the cornea, which is probably related to the specifics of corneal structure, or to its accessibility for examination.

3. Theoretical (mathematical) biomechanics

The originator of mathematical approach to study biomechanical properties of the cornea was F.A. Rachevsky. In 1930, in his theoretical study, he pointed "...for the first time at the paramount importance of the radius of the corneal curve and especially of its thickness for specific results of intraocular pressure tonometry." Besides that, he proved mathematically that under effect of external and internal forces, tangentially directed stress occurs in the cornea, particularly during applanation tonometry [4].

At present, the research of corneal biomechanics is conducted in two main directions. Mathematical modeling is generally used for the calculation of parameters and prediction of results of keratorefractive surgeries [5–8], as well as for the determination of possible procedural errors of applanation tonometry methods when biomechanical properties changed as the result of a surgery or a disease [9].

The main obstacle for proper mathematical modeling is anisotropy of the cornea. The majority of the proposed models does not consider it, which limits their application in practical ophthalmology [10–12].

According to Pinsky et al., the anisotropy of the cornea primarily depends on its structural features, that is, specific architectural organization of collagen fibers [13]. X-ray structural analysis revealed that collagen fibrils of the central area have orthogonal orientation predominantly in vertical and horizontal directions, while fibrils of the periphery have tangential orientation [14]. Pinsky et al. developed a mathematical model for corneal anisotropy mechanics that accounts for these findings [13]. Based on the finite element method, the model allows predicting biomechanical response of the cornea to tunnel cutting, radial keratotomy, and LASIK [15–17].

In order to determine the possible error margin of applanation tonometry methods, several mathematical models have been developed [18]. Liu et al. used mathematical modeling to study isolated effects of various biometric and biomechanical parameters of the cornea on Goldmann tonometry readings [19]. Kwon et al. developed a mathematical model demonstrating the need to take into account not only corneal thickness, but also its biomechanical properties when interpreting Goldmann tonometry readings [20].

4. Physical (experimental) biomechanics

4.1 Normal (intact) cornea

Experimental studies based on extensiometry revealed distinguishing biomechanical anisotropy and heterogeneity of the cornea. Corneal material acquired with a radial cut has the best durability and margin of deformation capacity. Those parameters decrease with distance from the radial direction. Corneal material stretched tangentially shows approximately the same elastic properties along the corneal disc. The samples stretched radially appeared to have the highest rigidity. In the course of the study, Poisson's ratio was determined for various parts of the cornea. This ratio characterizes the transverse deformation lateral to stretch direction, for radial direction, it was in the range of 0.445–0.450, and for tangential direction, it was in the range from 0.290 to 0.310 (middle periphery) and from 0.340 to 0.350 (perilimbal) [21, 22].

A variety of studies is dedicated to measuring the main elastic and strength properties of the cornea, but analysis of the data shows that isolated corneas exhibit big spread in the values—from 0.3 to 13.6 MPa. The phenomenon can be attributed to different experimental conditions and nonlinear nature of biomechanical properties of corneal material [23–26]. Andreassen et al. studied the biomechanics of corneal discs with extensiometry; the discs were taken from patients with keratoconus after they underwent penetrating keratoplasty. The study revealed significant decrease of mechanical strength properties in pathologically altered corneas [27].

Soergel et al. used dynamic mechanical spectroscopy to evaluate viscoelastic properties of the cornea in experimental environment. They found that elastic and shearing deformation depend on the hydration, time elapsed after death, and temperature of the tissue [28].

Wang et al. calculated Young's modulus by measuring the speed of ultrasound transmission through cadaver cornea and processing the data with Fourier analysis [29].

Like ultrasound spectroscopy, Brillouin microscopy can determine intrinsic viscoelastic properties decoupled from the structural information and applied pressure. In contrast, it can measure the local acoustic properties with much higher spatial resolution and sensitivity, and the measurement is performed optically without the need for acoustic transducers or physical contact with the cornea [30].

One of the techniques, holographic interferometry, is used to calculate Young's modulus. The method is to some degree similar to videokeratography, that is, holographic technologies are used to examine the changes in corneal surface. A study conducted on an intact bull's cornea showed Young's modulus being two orders lower than when measured in an experiment with corneal tissue samples. The authors summarized that localization and hydration level plays the primary roles during measurement. This method, however, is limited in terms of practical use due to requiring maximum permissible laser emission in order for the resulting images to be high quality [31].

4.2 Cornea after refractive surgery

Some studies showed significant increase of tangential elasticity of the cornea after it was incised with radial cuts (up to 46.5% with an incision depth of 0.6 mm), that is, in the direction of the lesser material rigidity [32]. In certain cases, the changes led to severe complications in the long-term postoperative period. Particularly, it manifested as a significant decrease of eyeball's resistance to trauma with potential disruption of corneal cicatrices and loss of membranes [33].

Luminescent polariscopy revealed that after radial keratotomy, the main mechanical strain fell on the middle periphery of the cornea, particularly on the bottom of keratotomic incisions. An increase of intracameral pressure (analogue to intraocular pressure) raises the strain on peripheral part of the cornea and off-loads its central part, which can cause hypermetropic shift in refraction [34].

However, with the appearance and widespread implementation of excimer laser technologies for correction of refraction errors, such risks have greatly decreased. It can be attributed to different mechanisms of corneal refraction change, that is, its thinning in the central area.

Experimental studies on biomechanical properties of the cornea after excimer laser intervention indicate that thinning of the cornea in 6.0-mm optic zone for more than 15–20% results in significant changes of its mechanical properties. In terms of clinical relevance, the most meaningful change appears to be the significant (mean 20%) decrease of breaking load for experimental samples in comparison to the control samples. Additionally, changes in deformation properties of the cornea after laser ablation should also be taken into consideration, which manifested as lowered amount of movement the punch had to do before the cornea broke in experimental eyes in comparison to the control subjects in average by 10.72% [35].

However, the mechanical properties of the data obtained using an isolated cornea cannot objectively reflect the parameters of the tissues in natural environment. Adequate information on the biomechanical state of the cornea can only be obtained from a living eye.

5. Clinical biomechanics

5.1 Normal (intact) cornea

Clinical studies on the biomechanical properties of the relatively healthy cornea have been conducted since the middle of the twentieth century, but those methods remained widely unused due to various reasons.

In 1937, Friedenwald suggested that rigidity coefficient could be calculated based on a logarithmic dependence between IOP changes and eye volume employing differential tonometry with Schiotz tonometer [36]. Friedenwald depicted the relation between pressure and volume in a coordinate system. As was shown by further clinical studies, the proposed coefficient strongly depends on the corneal curvature and thickness, as well as on the IOP level [37]. According to research results, the parameter suggested by Friedenwald—the rigidity coefficient—was inaccurate in eyes with deviations in biomechanics (thickness and curvature) from the norm. It was also strongly influenced by IOP.

In 1936, S. F. Kalfa proposed a method of elastotonometry, that is, differential tonometry with four Maklakov tonometers weighing 5, 7.5, 10, and 15 g. Connecting the dots marked on a coordinate system forms an elastotonometric curve, which appears ascending line. The difference in mm Hg between the starting and ending points of the curve, that is, between IOP value obtained using 5.0 and 15.0 g tonometers, is called elasto-ascent. Essentially, Friedenwald's rigidity coefficient and S. F. Kalfa's elasto-ascent are different expressions of the same thing. In norm, the two figures are closely related, albeit not functionally [38].

There are a number of techniques described by their authors as potential intravital methods for examination of biomechanical properties of the cornea, but they have not been adopted into clinical practice: electronic speckle interferometry [39], dynamic cornea visualization [40], corneal applanation and indentation [41], ultrasound elastometry [42], and photoelasticity method [43].

As an alternative to holographic interferometry, a noncontact, nondisruptive method of electronic speckle interferometry was suggested; it is equally sensitive because it employs close wavelength for measurement. Advantages of the method include the absence of requirement of photographic hologram recording, which simplifies the procedure and enables real-time acquisition of corneal surface shift data using a television camera. The method is recommended for evaluation of changes in corneal biomechanics after excimer laser refractive surgery [39].

Grabner et al. proposed a method of dynamic visualization of the cornea. It involves applying dosed pressure to the central area of the cornea during videokeratography by means of a special indenter and subsequent analysis of the topographic pattern. As the result, high correlation between the bending curve and depression depth was found. The form of the curve was noted to be affected by central corneal thickness, intraocular pressure, and patient age. Moreover, bending curves were different in keratoconus patients, as well as in patients who had underwent keratorefractive surgeries [40].

Chang et al. studied biomechanical properties of the cornea *in vivo* using corneal applanation and indentation on rabbit and human eyes, regarding the cornea as a transversely isotropic material. The study showed normal Young's modulus to vary from 1 to 5 MPa and transverse shift modulus from 10 to 30 KPa [41].

Some authors used photoelasticity method to evaluate mechanical stress in the cornea involving the measurement of its polarization and optical properties [43].

Scoping a large amount of clinical data, Edmund calculated Young's modulus adhering to the hypothesis that the final form of cornea is the outcome of counteraction between tissue elasticity and intraocular pressure. The modulus values were significantly lower in keratoconus eyes when compared to norm. The study also showed significant difference between healthy and ectatic patients in relative distribution of stress in the central and peripheral areas of the cornea, which can help with the understanding of keratoconus pathogenesis. However, this method generally does not find much use in clinical practice [44].

The one method most widely used in present day clinical practice involves ocular response analyzer (ORA)—a device that analyses corneal biomechanical properties based on bi-directional corneal applanation by an air pulse [45]. The method's authors proposed to evaluate biomechanical response of the cornea by quantifying the differential inward and outward corneal response to an air pulse and thus obtaining two parameters—corneal hysteresis (CH) and corneal resistance factor (CRF). Corneal hysteresis characterizes the viscoelastic properties of the cornea responsible for the partial absorption of the air pulse energy. Corneal resistance factor is a derived parameter with high correlation to central corneal thickness that reflects the elastic properties of the cornea.

Multiple studies have confirmed the usefulness of bi-directional corneal applanation for the evaluation of biomechanical properties of the cornea: they rise with the increase of the corneal thickness [46, 47]. Corneal hysteresis was in the average 10.8 ± 1.5 mm Hg and corneal resistance factor—11.0 ± 1.6 mm Hg. Statistically, a significant difference in the mean values of CH and CRF between groups of varying age was absent, with the exception of patients older than 60 years for whom the values were on lower. It is possible that the phenomenon reflects the changes in elastic properties of the cornea associated with age, but the authors also note the potential influence of other parameters (intraocular pressure and anterior-posterior axis length) that were disregarded in the study [48]. The comparison of CH and CRF in children and adults did not reveal any age-related differences [49].

Studying the diurnal variations in CH and CRF parameters revealed their hourly stability, while minor changes observed between the morning and evening measurements can be explained by diurnal IOP fluctuations [50]. CH and CRF correlated strongly with corneal thickness and to a lesser degree with an amount of astigmatism. No correlation was found with keratometry, age, gender, spherical equivalent, or IOPcc [51]. Moreover, ORA shows good repeatability of biomechanical and tonometry measurements [52].

Avetisov et al. studied the possibility of applying the dynamic pneumo-impression of the cornea approach to the existing corneal biomechanical properties analyzer (ORA). The fundamental principle was that at the curvature start point laying on the border of the impression area, the pneumatic jet is subject to the counter-force of IOP and corneal elasticity, in equal amounts. At the moment of maximum impression, the pneumatic jet is mainly countered by corneal elasticity—due to the maximum deformation of the cornea. As the result, a parameter named elasticity coefficient was calculated characterizing the elastic behavior of the cornea regardless of the IOP level [53].

The same principle was used in CorVis device (Oculus, Germany), in which corneal deformation responding to a pulse of air is monitored with high-speed Scheimpflug camera. The device can help to measure a whole range of parameters that characterize the particularities of corneal deformation during the impression process. It records the process between the initial and the second applanations involving the cornea recovering its initial form, captures the maximum indentation point, and measures IOP [54, 55].

5.2 Keratoconic cornea

Intravital measurement of biomechanical properties of the cornea in keratoconus patients performed with dynamic bi-directional pneumo-applanation showed lower CH and CRF values than in healthy eyes. Apparent negative correlation between the CH and CRF parameters and the degree of keratoconus were also evident [56].

Additionally, CH was significantly higher than CRF in the keratoconus group. The authors suggested the CH decrease of less than 8 mm Hg in conjunction with positive CH-CRF difference to be considered a stronger sign of keratoconus than isolated decrease of CH. Glaucoma patients showed reverse tendency: CRF value was higher than CH [57].

Studying the parameters obtained with dynamic Scheimpflug analysis (Covis ST) showed the possibilities of the examination method for differential diagnostics of patients suspected of keratoconus or with early keratoconus from patients with normal cornea [58].

5.3 Cornea after refractive surgery

Intravital measurements of biomechanical properties of the cornea after excimer laser surgery performed using dynamic bi-directional pneumo-applanation also confirmed the loss of corneal strength. In patients who had underwent LASIK, examination showed decrease of IOP-related parameters such as corneal compensated IOP, as well as parameters reflecting the biomechanical properties. Along with that, significant correlation was observed between the amount of myopia correction and the deterioration of the biomechanical properties [59].

Another study analyzed the results of dynamic bi-directional pneumo-applanation of the cornea and assessed the correlation between CH decrease and ablation depth in three patient groups: after photorefractive keratectomy, after LASIK with mechanical corneal flap creation, and after LASIK with femtosecond flap creation. The authors found that the strongest correlation was present in femto-LASIK group, while in the two other groups, it was significantly lower [60].

Isolated creation of corneal flap was also found to cause minor changes in corneal refraction [61]. Roberts explained the phenomenon with a theory stating that after lamellar dissection, the corneal biomechanics change in such a way so that severed fibrils contract causing traction in the direction of limbus. With that, central corneal area deflates under the action of released fibrils inducing the so-called "hypermetropic" shift [62, 63].

In parallel, a comparison of changes in biomechanical properties of the cornea after superficial and intrastromal keratectomy was done using OCULUS Corvis (Germany) tonometer. Both types of keratectomy were found to cause statistically significant decrease of biomechanical parameters [64].

Despite the existing methods of measuring biomechanical properties of the cornea and the developed biomechanical models, the detection of ectasia after excimer laser vision correction varies from 0.04 to 0.6% of cases, but according to some researchers, the numbers may be an underestimation [65, 66].

Iatrogenic keratectasia is known to develop due to two factors: an ectatic corneal disease that was undiagnosed in the preoperative stage and excessive thinning of the cornea [67]. In the first case, early detection of keratoconus poses objective challenges [68, 69].

At the same time, even when keratoconus was timely diagnosed, the selection of candidates for keratorefractive surgeries is still difficult, and the evaluation of corneal biomechanics by means of dynamic bi-directional pneumo-applanation does not yield the necessary data.

6. Correction of corneal biomechanical properties

Presently, the most common method of correcting (strengthening) biomechanical properties of the cornea is corneal cross-linking [70].

The first specialists who in the 90s of the twentieth century created corneal cross-linking method for treating keratoconus were Wollensak, Spoerl, and Seiler [71]. They developed the protocol ("Dresden protocol") for using this method of strengthening the cross-link bonds of collagen for treating progressive keratoconus involving riboflavin and ultraviolet A irradiation of the corneal stroma (UVA with a wavelength of 370 nm for peak absorption of riboflavin) [72].

Careful preclinical experimental validation showed that the combination of riboflavin and UVA leads to a significant improvement of biomechanical stability of the cornea (increase of elastic modulus approximately by 300%) and the formation of large collagen molecular aggregates, including the appearance of cross links—predominantly between the fibril surface molecules and also between proteoglycans in the interfibrillar space [73–75].

In the following decades, the corneal cross-linking technique has seen widespread clinical application with indications for its usage expanding significantly. Effectiveness of the method for strengthening biomechanical properties was confirmed for the treatment of not only progressive keratoconus, but also pellucid marginal degeneration and iatrogenic ectasia caused by excimer laser surgery [76].

An important suggestion has been made recently for reinforcing the effect of corneal cross-linking—to combine the procedure with implantation of corneal segments [77, 78]. Comparative studies of different treatments—individually and in combination—showed the most pronounced effect to be from the combination therapy starting with the implantation of corneal segments and followed by cross-linking, and not in the reverse order. Such combination therapy also helps to achieve better results (weakening of manifest refraction and keratometric indicators) in cases with keratectasia after excimer laser surgery [79].

There is another method described in the literature as directed laser ablation; it involves biomechanical approach to ablation calculations. Vaporization of the tissue thus happens on the middle periphery, which contains certain relatively flat spots, and not in the central (thin) area. The rationale is that thinning of the area leads to steepening of the cornea subsequently flattening the unablated area, which has more optical power [80]. It should be noted that in clinical practice, this method requires very careful consideration and cautiousness due to insufficient studies on its after effects.

Furthermore, a multimodal approach involving implantation of intrastromal rings, CXL, and laser ablation in different configurations may provide not only stability of corneal topography, but also positive refraction result, thanks to the combination of all the methods' advantages [81–85]. That said, the lack of established standards and clinical recommendations for combining different methods for the correction of corneal biomechanical properties may lead to various complications and unexpected aftermaths; it should be kept in mind when planning such treatment.

7. Conclusion

In summary, clinical relevance of studying biomechanics of the fibrous tunic is difficult to overestimate. The diversity of methods used for examination of biomechanical properties of the cornea means there is no single method that could fully satisfy the needs of practical ophthalmology. Further studies are necessary to develop simple, available, and sufficiently informative method for clinical assessment of ocular biomechanics. Moreover, the demand for techniques of correcting biomechanical properties keeps growing, and so this field of research has wide potential.

Conflict of interest

The author has no conflict of interest.

Author details

Irina Bubnova
Research Institute of Eye Diseases, Moscow, Russian Federation

*Address all correspondence to: bubnova.irina@gmail.com

References

[1] Iomdina EN, Bauer SM, Kotliar KE. Eye Biomechanics: Theoretical Aspects and Clinical Applications. Moscow: Real Time; 2015. p. 208

[2] DelMonte DW, Kim T. Anatomy and physiology of the cornea. Journal of Cataract & Refractive Surgery. 2011;**37**(3):588-598

[3] Egorova E, Borsenok S, Bessarabov A, Miligert A, Sevostyanov M, Baikin A. Biomechanical properties of sclera in groups of patients with different refraction. Ophthalmosurgery. 2015;**4**:65-69

[4] Avetisov SE, Bubnova IA, Antonov AA. Corneal biomechanics: Clinical importance, evaluation, possibilities of sistemization of examination approaches. Vestnik Oftalmologii. 2010;**126**(6):3-7

[5] Lanchares E, Calvo B, Cristóbal JA, Doblaré M. Finite element simulation of arcuates for astigmatism correction. Journal of Biomechanics. 2008;**41**(4):797-805. DOI: 10.1016/j.jbiomech.2007.11.010

[6] Roy AS, Dupps WJ. Patient-specific modeling of corneal refractive surgery outcomes and inverse estimation of elastic property changes. Journal of Biomechanical Engineering. 2011;**133**(1):011002

[7] Achiron A, Gur Z, Aviv U, Hilely A, Mimouni M, Karmona L, et al. Predicting refractive surgery outcome: Machine learning approach with big data. Journal of Refractive Surgery. 2017;**33**(9):592-597. DOI: 10.3928/1081597X-20170616-03

[8] Bao F, Wang J, Cao S, Liao N, Shu B, Zhao Y, et al. Development and clinical verification of numerical simulation for laser in situ keratomileusis. Journal of the Mechanical Behavior of Biomedical Materials. 2018;**83**:126-134. DOI: 10.1016/j.jmbbm.2018.04.016

[9] Bauer SM, Venatovskaya LA, Voronkova EB, Smirnov AL. The three-dimensional problem of the axisymmetric deformation of an orthotropic spherical layer. Vestnik St Petersburg University: Mathematics. 2016;**49**(3):277-283

[10] Pandolfi A, Manganiello F. A model for the human cornea: Constitutive formulation and numerical analysis. Biomechanics and Modeling in Mechanobiology. 2006;**5**(4):237-246. DOI: 10.1007/s10237-005-0014-x

[11] Bauer S, Venatovskaya L, Franus D, Fedotova L. Estimation of changes in the stress-strain state of an eye shell and intraocular pressure readings after refractive correction of hyperopia. Russian Journal of Biomechanics/Rossijski Zurnal Biomehaniki. 2015;**19**(2):136-143

[12] Bryant MR, McDonnell PJ. Constitutive laws for biomechanical modeling of refractive surgery. Journal of Biomechanical Engineering. 1996;**118**(4):473-481

[13] Pinsky PM, van der Heide D, Chernyak D. Computational modeling of mechanical anisotropy in the cornea and sclera. Journal of Cataract & Refractive Surgery. 2005;**31**(1):136-145. DOI: 10.1016/j.jcrs.2004.10.048

[14] Meek KM, Newton RH. Organization of collagen fibrils in the corneal stroma in relation to mechanical properties and surgical practice. Journal of Refractive Surgery. 1999;**15**(6):695-699

[15] Roy AS, Dupps WJ. Effects of altered corneal stiffness on native and postoperative LASIK corneal biomechanical behavior: A whole-eye

finite element analysis. Journal of Refractive Surgery. 2009;**25**(10):875-887

[16] Kling S, Marcos S. Finite-element modeling of intrastromal ring segment implantation into a hyperelastic cornea. Investigative Ophthalmology & Visual Science. 2013;**54**(1):881-889

[17] Choi K-S, Soo S, Chung F-L. A virtual training simulator for learning cataract surgery with phacoemulsification. Computers in Biology and Medicine. 2009;**39**(11):1020-1031

[18] Elsheikh A, Wang D, Kotecha A, Brown M, Garway-Heath D. Evaluation of Goldmann applanation tonometry using a nonlinear finite element ocular model. Annals of Biomedical Engineering. 2006;**34**(10):1628-1640

[19] Liu J, Roberts CJ. Influence of corneal biomechanical properties on intraocular pressure measurement: Quantitative analysis. Journal of Cataract & Refractive Surgery. 2005;**31**(1):146-155. DOI: 10.1016/j.jcrs.2004.09.031

[20] Kwon TH, Ghaboussi J, Pecknold DA, Hashash YM. Role of corneal biomechanical properties in applanation tonometry measurements. Journal of Refractive Surgery. 2010;**26**(7):512-519

[21] Elsheikh A, Anderson K. Comparative study of corneal strip extensometry and inflation tests. Journal of the Royal Society Interface. 2005;**2**(3):177-185

[22] Avetisov SE, Mamikonyan VR, Zavalishin NN, Nenukov AK. Experimental study of mechanical properties of cornea and sclera. Ophthalmic Journal. 1988;**4**:233-237

[23] Elsheikh A, Brown M, Alhasso D, Rama P, Campanelli M, Garway-Heath D. Experimental assessment of corneal anisotropy. Journal of Refractive Surgery. 2008;**24**(2):178-187

[24] Shin TJ, Vito RP, Johnson LW, McCarey BE. The distribution of strain in the human cornea. Journal of Biomechanics. 1997;**30**(5):497-503

[25] Zeng Y, Yang J, Huang K, Lee Z, Lee X. A comparison of biomechanical properties between human and porcine cornea. Journal of Biomechanics. 2001;**34**(4):533-537

[26] Woo S-Y, Kobayashi A, Schlegel W, Lawrence C. Nonlinear material properties of intact cornea and sclera. Experimental Eye Research. 1972;**14**(1):29-39

[27] Andreassen TT, Simonsen AH, Oxlund H. Biomechanical properties of keratoconus and normal corneas. Experimental Eye Research. 1980;**31**(4):435-441

[28] Soergel F, Jean B, Seiler T, Bende T, Mücke S, Pechhold W, et al. Dynamic mechanical spectroscopy of the cornea for measurement of its viscoelastic properties in vitro. German Journal of Ophthalmology. 1995;**4**(3):151-156

[29] Wang H, Prendiville PL, McDonnell PJ, Chang WV. An ultrasonic technique for the measurement of the elastic moduli of human cornea. Journal of Biomechanics. 1996;**29**(12):1633-1636

[30] Scarcelli G, Pineda R, Yun SH. Brillouin optical microscopy for corneal biomechanics. Investigative Ophthalmology & Visual Science. 2012;**53**(1):185-190

[31] Kasprzak H, FÖrster W. Measurement of elastic modulus of the bovine cornea by means of holographic interferometry. Part 1. Method and experiment. Optometry and Vision Science. 1993;**70**(7):535-544

[32] Avetisov SE, Fyodorov AA, Vvedenskyi AS, Nenukov AK. Experimental study of radial keratotomy influence on corneal

mechanical properties. Ophthalmic Journal. 1990;**1**:54-58

[33] Forstot SL, Damiano RE. Trauma after radial keratotomy. Ophthalmology. 1988;**95**(6):833-835

[34] Avetisov S, Bubnova I, Novikov I, Antonov A, Siplivyi V. Experimental study on the mechanical strain of corneal collagen. Journal of Biomechanics. 2013;**46**(10):1648-1654. DOI: 10.1016/j.jbiomech.2013.04.008

[35] Avetisov SE, Voronin GV. Experimental study of corneal mechanical properties after eximer laser ablation. RMJ Clinical Ophthalmology. 2001;**3**:83-86

[36] Friedenwald JS, Moses R. Modern refinements in tonometry. Documenta Ophthalmologica. 1950;**4**(1):335-362

[37] Moses RA. Theory of the Schiotz tonometer and its empirical calibration. Transactions of the American Ophthalmological Society. 1971;**69**:494

[38] Avetisov SE, Bubnova IA, Antonov AA. Once more about the diagnostic capacities of elastic tonometry. Vestnik Oftalmologii. 2008;**124**(5):19-21

[39] Jaycock PD, Lobo L, Ibrahim J, Tyrer J, Marshall J. Interferometric technique to measure biomechanical changes in the cornea induced by refractive surgery. Journal of Cataract & Refractive Surgery. 2005;**31**(1):175-184

[40] Grabner G, Eilmsteiner R, Steindl C, Ruckhofer J, Mattioli R, Husinsky W. Dynamic corneal imaging. Journal of Cataract & Refractive Surgery. 2005;**31**(1):163-174

[41] Chang S, Hjortdal J, Maurice D, Pinsky P. Corneal Deformation by Indentation and Applanation Forces. Investigative Ophthalmology & Visual Science. Philadelphia, PA: Lippincott-Raven Publ; 1993. p. 19106

[42] Dupps WJ, Netto MV, Herekar S. Surface wave elastometry of the cornea in porcine and human donor eyes. Journal of Refractive Surgery. 2007;**23**(1):66-75

[43] Volokov V, Juravlev A, Malyshev L, Saulgosis U, Nekrasov U, Pavilainen V. Current state and prospects of application of the method of photoelasticity in ophthalmology. Journal of Ophthalmology. 1990;**8**:479-482

[44] Edmund C. Corneal topography and elasticity in normal and keratoconic eyes. A methodological study concerning the pathogenesis of keratoconus. Acta Ophthalmologica. Supplement. 1989;**193**:1-36

[45] Luce DA. Determining in vivo biomechanical properties of the cornea with an ocular response analyzer. Journal of Cataract & Refractive Surgery. 2005;**31**(1):156-162. DOI: 10.1016/j.jcrs.2004.10.044

[46] Medeiros FA, Weinreb RN. Evaluation of the influence of corneal biomechanical properties on intraocular pressure measurements using the ocular response analyzer. Journal of Glaucoma. 2006;**15**(5):364-370

[47] Annette H, Kristina L, Bernd S, Mark-Oliver F, Wolfgang W. Effect of central corneal thickness and corneal hysteresis on tonometry as measured by dynamic contour tonometry, ocular response analyzer, and Goldmann tonometry in glaucomatous eyes. Journal of Glaucoma. 2008;**17**(5):361-365

[48] Kotecha A, Elsheikh A, Roberts CR, Zhu H, Garway-Heath DF. Corneal thickness-and age-related biomechanical properties of the cornea measured with the ocular response analyzer. Investigative Ophthalmology & Visual Science. 2006;**47**(12):5337-5347

[49] Kirwan C, O'keefe M, Lanigan B. Corneal hysteresis and intraocular pressure measurement in children using the reichert ocular response analyzer. American Journal of Ophthalmology. 2006;**142**(6):990-992

[50] Gonzalez-Meijome JM, QueirOS A, Jorge J, DIAz-Rey A, Parafita MA. Intraoffice variability of corneal biomechanical parameters and intraocular pressure (IOP). Optometry and Vision Science. 2008;**85**(6):457-462

[51] Montard R, Kopito R, Touzeau O, Allouch C, Letaief I, Borderie V, et al. Ocular response analyzer: Feasibility study and correlation with normal eyes. Journal Français d'Ophtalmologie. 2007;**30**(10):978-984

[52] Moreno-Montanés J, Maldonado MJ, García N, Mendiluce L, García-Gómez PJ, Seguí-Gómez M. Reproducibility and clinical relevance of the ocular response analyzer in nonoperated eyes: Corneal biomechanical and tonometric implications. Investigative Ophthalmology & Visual Science. 2008;**49**(3):968-974

[53] Avetisov SE, Novikov IA, Bubnova IA, Antonov AA, Siplivyi VI. Determination of corneal elasticity coefficient using the ORA database. Journal of Refractive Surgery. 2010;**26**(7):520-524

[54] Reznicek L, Muth D, Kampik A, Neubauer AS, Hirneiss C. Evaluation of a novel Scheimpflug-based non-contact tonometer in healthy subjects and patients with ocular hypertension and glaucoma. British Journal of Ophthalmology. 2013;**97**(11):1410-1414

[55] Asaoka R, Nakakura S, Tabuchi H, Murata H, Nakao Y, Ihara N, et al. The relationship between Corvis ST tonometry measured corneal parameters and intraocular pressure, corneal thickness and corneal curvature. PLoS One. 2015;**10**(10):e0140385

[56] Fontes BM, Ambrósio R, Velarde GC, Nosé W. Ocular response analyzer measurements in keratoconus with normal central corneal thickness compared with matched normal control eyes. Journal of Refractive Surgery. 2011;**27**(3):209-215

[57] Touboul D, Roberts C, Kérautret J, Garra C, Maurice-Tison S, Saubusse E, et al. Correlations between corneal hysteresis, intraocular pressure, and corneal central pachymetry. Journal of Cataract & Refractive Surgery. 2008;**34**(4):616-622

[58] Francis M, Pahuja N, Shroff R, Gowda R, Matalia H, Shetty R, et al. Waveform analysis of deformation amplitude and deflection amplitude in normal, suspect, and keratoconic eyes. Journal of Cataract & Refractive Surgery. 2017;**43**(10):1271-1280

[59] Pedersen IB, Bak-Nielsen S, Vestergaard AH, Ivarsen A, Hjortdal J. Corneal biomechanical properties after LASIK, ReLEx flex, and ReLEx smile by Scheimpflug-based dynamic tonometry. Graefe's Archive for Clinical and Experimental Ophthalmology. 2014;**252**(8):1329-1335

[60] Hamilton DR, Johnson RD, Lee N, Bourla N. Differences in the corneal biomechanical effects of surface ablation compared with laser in situ keratomileusis using a microkeratome or femtosecond laser. Journal of Cataract & Refractive Surgery. 2008;**34**(12):2049-2056

[61] Güell JL, Velasco F, Roberts C, Sisquella MT, Mahmoud A. Corneal flap thickness and topography changes induced by flap creation during laser in situ keratomileusis. Journal of Cataract & Refractive Surgery. 2005;**31**(1): 115-119. DOI: 10.1016/j.jcrs.2004.09.045

[62] Potgieter FJ, Roberts C, Cox IG, Mahmoud AM, Herderick EE, Roetz M, et al. Prediction of flap response.

Journal of Cataract & Refractive Surgery. 2005;**31**(1):106-114

[63] Roberts C. Biomechanical customization: The next generation of laser refractive surgery. Journal of Cataract & Refractive Surgery. 2005;**31**(1):2-5

[64] Hassan Z, Modis L, Szalai E, Berta A, Nemeth G. Examination of ocular biomechanics with a new Scheimpflug technology after corneal refractive surgery. Contact Lens & Anterior Eye. 2014;**37**(5):337-341. DOI: 10.1016/j.clae.2014.05.001

[65] Vahdati A, Seven I, Mysore N, Randleman JB, Dupps WJ. Computational biomechanical analysis of asymmetric ectasia risk in unilateral post-LASIK ectasia. Journal of Refractive Surgery. 2016;**32**(12) :811-820. DOI: 10.3928/1081597X-20160929-01

[66] Santhiago MR, Giacomin NT, Smadja D, Bechara SJ. Ectasia risk factors in refractive surgery. Clinical Ophthalmology (Auckland, NZ). 2016;**10**:713. DOI: 10.2147/OPTH.S51313

[67] Pallikaris IG, Kymionis GD, Astyrakakis NI. Corneal ectasia induced by laser in situ keratomileusis. Journal of Cataract & Refractive Surgery. 2001;**27**(11):1796-1802

[68] Comaish IF, Lawless MA. Progressive post-LASIK keratectasia: Biomechanical instability or chronic disease process? Journal of Cataract & Refractive Surgery. 2002;**28**(12):2206-2213

[69] Dupps WJ. Biomechanical modeling of corneal ectasia. Journal of Refractive Surgery. 2007;**23**(1):186-190

[70] Roberts CJ, Dupps WJ. Biomechanics of corneal ectasia and biomechanical treatments. Journal of Cataract & Refractive Surgery. 2014;**40**(6):991-998. DOI: 10.1016/j.jcrs.2014.04.013

[71] Spoerl E, Seiler T. Techniques for stiffening the cornea. Journal of Refractive Surgery. 1999;**15**(6):711-713

[72] Spoerl E, Mrochen M, Sliney D, Trokel S, Seiler T. Safety of UVA-riboflavin cross-linking of the cornea. Cornea. 2007;**26**(4):385-389

[73] Wollensak G, Spoerl E, Seiler T. Stress-strain measurements of human and porcine corneas after riboflavin–ultraviolet-A-induced cross-linking. Journal of Cataract & Refractive Surgery. 2003;**29**(9):1780-1785

[74] Kohlhaas M, Spoerl E, Schilde T, Unger G, Wittig C, Pillunat LE. Biomechanical evidence of the distribution of cross-links in corneastreated with riboflavin and ultraviolet A light. Journal of Cataract & Refractive Surgery. 2006;**32**(2):279-283

[75] Seiler T, Hafezi F. Corneal cross-linking-induced stromal demarcation line. Cornea. 2006;**25**(9):1057-1059

[76] Hersh PS, Greenstein SA, Fry KL. Corneal collagen crosslinking for keratoconus and corneal ectasia: One-year results. Journal of Cataract & Refractive Surgery. 2011;**37**(1):149-160. DOI: 10.1016/j.jcrs.2010.07.030

[77] Dauwe C, Touboul D, Roberts CJ, Mahmoud AM, Kérautret J, Fournier P, et al. Biomechanical and morphological corneal response to placement of intrastromal corneal ring segments for keratoconus. Journal of Cataract & Refractive Surgery. 2009;**35**(10): 1761-1767. DOI: 10.1016/j.jcrs.2009.05.033

[78] Kılıç A, Kamburoglu G, Akıncı A.Riboflavin injection into the corneal channel for combined collagen crosslinking and intrastromal corneal

ring segment implantation. Journal of Cataract & Refractive Surgery. 2012;**38**(5):878-883

[79] Coskunseven E, Jankov MR II, Hafezi F, Atun S, Arslan E, Kymionis GD. Effect of treatment sequence in combined intrastromal corneal rings and corneal collagen crosslinking for keratoconus. Journal of Cataract & Refractive Surgery. 2009;**35**(12):2084-2091

[80] Cennamo G, Intravaja A, Boccuzzi D, Marotta G, Cennamo G. Treatment of keratoconus by topography-guided customized photorefractive keratectomy: Two-year follow-up study. Journal of Refractive Surgery. 2008;**24**(2):145-149

[81] Stojanovic A, Zhang J, Chen X, Nitter TA, Chen S, Wang Q. Topography-guided transepithelial surface ablation followed by corneal collagen cross-linking performed in a single combined procedure. Journal of Refractive Surgery. 2010;**26**(2):145-152. DOI: 10.3928/1081597X-20100121-10

[82] Krueger RR, Kanellopoulos AJ. Stability of simultaneous topography-guided photorefractive keratectomy and riboflavin/UVA cross-linking for progressive keratoconus. Journal of Refractive Surgery. 2010;**26**(10):S827-SS32. DOI: 10.3928/1081597X-20100921-11

[83] Kamburoglu G, Ertan A. Intacs implantation with sequential collagen cross-linking treatment in postoperative LASIK ectasia. Journal of Refractive Surgery. 2008;**24**(7):S726-S7S9

[84] Kanellopoulos AJ. Comparison of sequential vs same-day simultaneous collagen cross-linking and topography-guided PRK for treatment of keratoconus. Journal of Refractive Surgery. 2009;**25**(9):S812-S8S8. DOI: 10.3928/1081597X-20090813-10

[85] Kymionis GD, Kontadakis GA, Kounis GA, Portaliou DM, Karavitaki AE, Magarakis M, et al. Simultaneous topography-guided PRK followed by corneal collagen cross-linking for keratoconus. Journal of Refractive Surgery. 2009;**25**(9):S807-SS11. DOI: 10.3928/1081597X-20090813-09

Section 5

Biomechanics of Connective Tissues

Chapter 5

Multi-Scale Biomechanics of Osteoporotic Spine Fracture

Haisheng Yang

Abstract

Osteoporosis, the most common bone disorder found in the elderly, afflicts from 15 to 30% Caucasian women in US and results in an estimated 700,000 spine fractures per year. The prevalence of spine fractures in the elderly is high, ranging from 20 to 50%. Fractures are biomechanical events resulting from the load applied to a bone exceeding its ability to bear load. Osteoporotic spine fracture occurs owing to diminished vertebral microarchitecture and microfailure of bone tissues, ultimately leading to a compromised whole vertebral strength, and therefore, it is a multi-scale biomechanics event. In this chapter, insights into the micromechanics of the human vertebral body gained by micro-computed tomography (CT) and micro-finite element modeling will be reviewed. Following that, noninvasive assessment of vertebral strength using quantitative CT-based finite element analysis at a continuum level and its potential applications in improving spine fracture risk prediction in the clinic will be discussed.

Keywords: osteoporosis, vertebral fracture, micromechanics, computed tomography (CT), finite element analysis, bone strength

1. Introduction

1.1 Osteoporosis and osteoporotic spine fracture

Osteoporosis is a metabolic disease characterized by an imbalance in bone formation and resorption that results in accelerated bone loss and deterioration of bone microarchitecture. This low bone mass and deteriorated microarchitecture cause a reduction in bone strength and an increased risk of fracture. Osteoporosis is the most common bone disorder found in the elderly [1]. The vertebral body and femur are common locations for osteoporosis-related fractures.

Osteoporosis results in an estimated 700,000 spine fractures per year. The prevalence of vertebral fractures in the elderly ranges from 20 to 50% with the higher prevalence in older age groups [2]. Women are more affected than men, although at least one study has reported equal prevalence of vertebral fractures in women and men [3]. Reports from the European vertebral osteoporosis study suggest that the prevalence of vertebral fractures in men is 21% at age 60–64 and increases to 29.1% at age 75–79 [4], an increase in prevalence of fracture of about 70% per decade, the same as is found in women. A National Osteoporosis Foundation expert panel estimated that 90% of vertebral fractures in white men were attributable to osteoporosis, as compared to 75% in men of other ethnic groups [5]. Thus, age-related vertebral fractures are widespread for both sexes, and as the size of the aging population continues to increase, their incidence is expected to increase.

In particular, by the year 2050, there will be nearly five times as many people over 85 in the US as there were in 1980 and 22% of the population will be over 65 compared to only 4% in 1900 [6]. These demographic trends make the need to reduce morbidity among elderly men and women an urgent priority. The problem is also a global one. Worldwide, osteoporotic fractures are expected to increase greatly, and the associated costs may have a devastating effect on the under-funded healthcare systems of many countries [1].

Given the clinical importance of osteoporosis, it is critical to accurately identify individuals who are at risk of fracture so that treatments can be taken to prevent fractures. According to the World Health Organization (WHO) standard, osteoporosis is presently identified by bone mineral density (BMD) measurement by dual energy X-ray absorptiometry (DXA). An individual is diagnosed as osteoporosis when his/her DXA-measured t-score is less than 2.5 standard deviations of that for young women [7]. The problem is that DXA works less successfully for predicting vertebral fractures [8]. BMD alone has difficulties distinguishing between patients with and without vertebral fractures. A previous study showed that, based on DXA measured t-scores, only 44% of women and 21% of men who had nonvertebral osteoporotic fractures were diagnosed as osteoporosis [9]. This observation suggests that more than half of those individuals who eventually experience fractures are not identified as osteoporotic based on BMD measurement. These high-risk individuals often do not get drug treatments, which can effectively reduce the risk of fracture. These results have indicated the need to develop the means of assessing fracture risk beyond the bone mineral density.

Development of improved methods to diagnose and monitor osteoporosis is a fundamental aspect of any strategy to both prevent and treat this disease. Toward that end, one major obstacle in improving vertebral fracture risk assessment is the poorly understood nature of the biomechanical mechanisms of vertebral strength and the etiology of vertebral fractures [10, 11], since the vertebral fracture is caused by deteriorated microarchitecture and microfailure of bone tissues leading to a compromised whole vertebral strength, and it is essentially a multi-scale biomechanical event.

1.2 Functional anatomy of the vertebra

The human vertebral column is composed of 33 vertebrae, including 7 cervical vertebrae, 12 thoracic vertebrae, 5 lumbar vertebrae, 5 sacrum vertebrae, and 4 coccyx vertebrae. The vertebrae are separated by intervertebral discs. Each vertebra has four main structural components, including the trabecular centrum, the superior and inferior endplates, the surrounding cortex, and the posterior elements (e.g., neural arch). The vertebral body primarily resists compressive forces acting down along the spinal column.

The endplates transmit loads between the vertebral body and the intervertebral disc. Also, the porous endplates function as a nutrient pathway between the disc and the vertebral body. The microstructure of the endplates is more like condensed trabeculae rather than compact cortical bone [12]. The thickness of the endplates is ~0.4–0.8 mm and varies across spinal level [13]. Generally, the endplates are thinner in the center than in the periphery. At certain spinal levels, inferior endplates are thicker than the superior endplates [14]. Endplates are common regions in the vertebral body suffering from osteoporotic fractures and have been thought to be "weak-link" of the lumbar spine [10].

The cortical shell is located at the periphery of the vertebral body and surrounds the trabecular bone inside the vertebral body. Although the cortical shell is only ~0.25–0.4 mm [15] in thickness and makes up only ~10–20% of the total amount of bone tissue in the vertebral body [16], it can carry up to 75% of the axial compressive load [17]. The cortex is thickest near the endplates and thinnest in the mid-transverse region.

The trabecular bone is located in the interior of the vertebral body. Vertebral trabecular bone has a highly porous (>80% porosity) plate- and rod-like architecture, which provides a unique spatial network to take and distribute loads effectively [18]. Aging- or osteoporosis-associated deterioration of the trabecular microarchitecture can lead to a reduction in whole vertebral strength and an increase in fracture risk. Trabecular microarchitecture can be characterized by bone volume fraction (bone tissue volume/total volume) and other microarchitecture parameters (e.g., trabecular thickness, trabecular number, trabecular separation, structural model index, connectivity density, and degree of anisotropy) that refer to the structure, interconnection, and spatial organization of the trabeculae.

1.3 Biomechanics of vertebral fractures

Vertebral compression fractures can be categorized into anterior wedge fracture, biconcave fracture, and crush fracture, based on their deformities. Anterior wedge fracture is the most common type of vertebral fracture, but its etiology remains unclear. Many osteoporotic vertebral fractures occur due to nontraumatic loading conditions, whereas hip fractures are attributable to a fall in approximately 90% of all cases [19]. The nontraumatic nature of vertebral fractures makes it difficult to diagnose because they are accompanied silently by microstructure deterioration and bone tissue failure.

Vertebral fractures are mechanical events that occur when the applied load exceeds the ability of the vertebral body to withstand load (i.e., vertebral strength). Based on this simple biomechanics concept, a factor of risk relevant to fracture can be defined as applied load over vertebral bone strength [20]. If the factor of risk exceeds 1, then factures are expected; if the factor of risk is less than 1, then the vertebral body is not expected to fracture. Apparently, the occurrence of a vertebral fracture depends on the mechanical loads acting on the vertebral body and, more importantly, the vertebral strength determined by its geometry, microarchitecture, bone tissue properties, etc.

Substantial changes of vertebra occur with aging and osteoporosis. Decreases in vertebral strength are caused primarily by the loss of bone density and deterioration in bone microarchitecture with age. One study estimated that the vertebral strength decreases by about 12% per decade from ages 25 to 85 [21]. Aging is also accompanied by the changes in the intervertebral disc, including disc degeneration. While age accounts for a large proportion of the variation in bone strength, individuals can show much stronger or weaker bones than would be predicted by their age alone. Similarly, BMD itself can account for some of the variations in bone strength but not all. At a given bone mineral density, the measured strength values for different individuals can be higher or lower than the expected value. Therefore, BMD measures cannot fully reflect bone strength that is directly related to fractures. This issue again highlights the importance of understanding vertebral fractures from a biomechanical perspective. On the one hand, we need an improved understanding of failure mechanisms of the vertebral body, particularly at a microlevel. For example, how do loads transfer from the intervertebral disc through the endplates into the trabecular and cortical bone and ultimately cause failure of vertebral microarchitecture? How does bone mechanical behavior at a tissue-level link to whole vertebral strength. On the other hand, a better understanding of the biomechanics of vertebral fractures can guide us to develop more advanced approaches for evaluating vertebral strength in clinic and predicting fracture risk other than BMD. Section 2 will describe some insights into the micromechanics of the human vertebral body, derived from large-scale microCT-based finite element analysis. Section 3 will discuss an improved method based on a combination of biomechanics and CT scans to predict vertebral fracture risk, which has been shown potentials in clinic use.

2. Micromechanics of the human vertebral body

2.1 MicroCT-based finite element analysis

Computed tomography (CT) is an imaging procedure in which a narrow beam of X-rays is sent out from an X-ray source, penetrating through a specimen or a patient's body, and detected by multiple detectors to generate computerized cross-sectional images. Since different materials have their own attenuation properties, CT images can indicate basic structures or anatomies by different gray scale values. MicroCT is a type of high-resolution CT with voxel sizes down to several micrometers. MicroCT has been widely used to reveal in great detail the internal microstructure of bone, particularly for trabecular bone. Due to its high-resolution nature, microCT generally requires a limited dimension of the scanned specimen and has usually been used in laboratory research. However, in vivo high-resolution imaging techniques, such as high-resolution peripheral quantitative CT (HR-pQCT) and high-resolution magnetic resonance (MR) imaging, have been developed and used in clinical studies [22]. However, those imaging techniques have not been seen for using in human spine.

Finite element analysis is a powerful computational tool that can be used to study bone biomechanics. A "virtually real" experiment can be performed by finite element analysis. This technique has some advantages over traditional biomechanical experiments. First, the technique is noninvasive, and it can be performed parametrically. For example, the effects of boundary and loading conditions as well as material properties can be examined in a controlled and repeated manner. Second, the finite element model can provide much detailed information on stress and strain distributions within the vertebra as well as mechanical behaviors of a whole vertebra, while only the apparent-level mechanical properties can be obtained from experimental testing. The greatest benefit of finite element modeling in bone mechanics research may be achieved by combining the technique with biomechanical testing to leverage the individual strengths of each approach.

Finite element models can be generated directly from micro-CT images (e.g., 10–60 μm voxel resolution) using a voxel conversion approach, with which voxels representing bone tissue are converted to eight-node brick elements, whereas voxels representing bone marrow and other tissues are ignored. Elements are typically assigned with homogeneous and isotropic material properties. By simulating a loading condition, this model can be used to determine the apparent-level mechanical properties of the bone (e.g., stiffness and strength) as well as tissue-level stress/strain in the bone. This modeling approach can be validated by comparing predicted outcomes with experimental measures. The microCT-based finite element analysis was first used in 1995 to investigate the mechanical behavior of trabecular bone [23]. Since then, much insight has been gained into the micromechanics of trabecular bone by using microCT-based finite element analysis [24, 25]. MicroCT-based finite element modeling of the human whole vertebral body was first reported in

2004 [26]. The models of whole vertebrae typically have in the order of 300 million degrees of freedom, and thus, the analyses require substantial parallel computational power, especially when geometric and material nonlinearities are considered.

The whole vertebral finite element models implicitly capture the trabecular microarchitecture, the thin cortical shell, and the porous endplates of the vertebra. Once the general modeling approach and model parameters have been validated, the models can be used to understand the micromechanics of the vertebral body and to link the microarchitecture or tissue material property alterations with whole vertebral mechanical behavior. Also, the microCT-based finite element models can be used to elucidate the failure mechanisms in the trabecular bone, cortical shell, and endplates. All these cannot be done with conventional experiments. So far, this technique has been applied to human whole vertebral bodies to determine cortical and trabecular load sharing, locations of high-risk tissue distributions, mechanisms of vertebral endplate failure, and relationship between microarchitecture and whole vertebral mechanical behavior [16, 26, 27]. The following subsections will introduce some additional detailed findings from the author's own work based on the microCT-finite element analysis of human whole vertebrae.

2.2 Effects of compression and forward flexion on the risk of vertebral failure

Uniform compression and forward flexion are common loading conditions that our vertebral bodies experience during daily life, corresponding to upright stance and stoop postures. Despite the high prevalence of vertebral wedge fractures in the clinical population, the mechanics of these fractures are not well understood. Clearly, the morphology of wedge fractures, in which the anterior side is shortened in the order of 15% more than the posterior side, is suggestive of an important role of forward flexion. Forward flexion of the spine motion segment might increase stresses within the anterior vertebral body including the cortical shell and trabecular bone, which could further cause failures of those bone tissues. Experiments using miniature pressure transducers have measured a greater pressure in the anterior half of the intervertebral disc when the disc-vertebra-disc segment was loaded in forward flexion [28]. However, whether this flexion-induced increase in the pressure of the anterior disc would increase the tissue stresses within the underlying cortical and trabecular microstructure of the vertebra remain unclear. This is mainly due to the technical difficulty of measuring such stresses in vivo or in vitro. Finite element modeling based on high-resolution microCT images of the vertebra is well suited to address this difficulty.

To gain insight into the etiology of wedge fractures, studies have been performed to investigate the high-risk tissue distribution within the human vertebral body for both forward flexion and uniform compression loading conditions [29]. Micro-CT-based high-resolution (60 μm) finite element models of 22 human T9 vertebral bodies with compliant discs (elastic modulus = 8 MPa) were built, and linear elastic finite element analysis was performed to mimic forward flexion or uniform compression loading. The compliant disc was also replaced with a stiff polymethylmethacrylate (PMMA, elastic modulus = 2500 MPa) layer in the vertebra-disc model to mimic a cadaveric experimental case. Results demonstrated that forward flexion increased the overall compressive load on the anterior half of the intervertebral disc. The spatial distribution of the vertebral bone tissues at the highest risk of initial failure, being identified as the top 10% percent of highly stressed bone tissues of the vertebral model, was shifted slightly toward the anterior aspect of the vertebral body. Despite that, the high-risk bone tissues were located primarily within the central regions of the trabecular bone and endplates (**Figure 1**). However, when the compliant disc

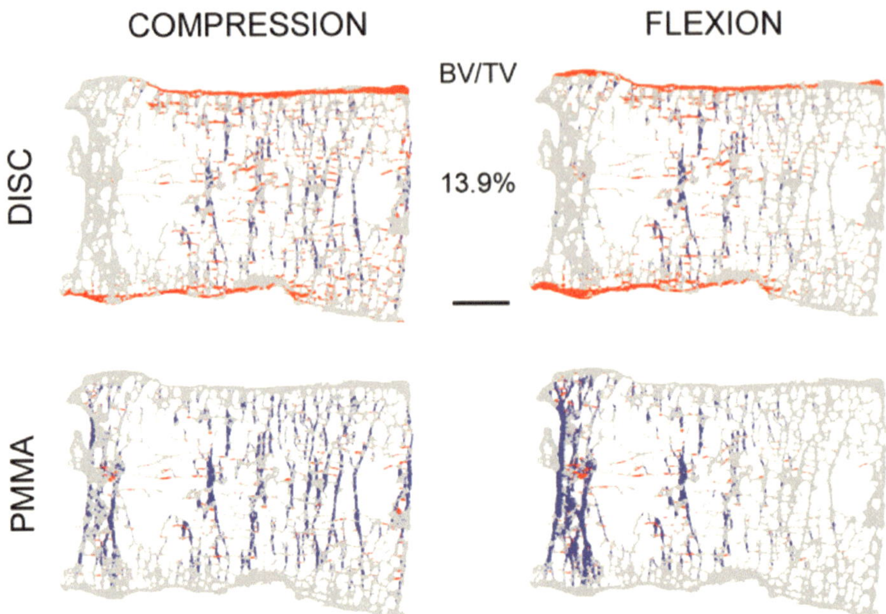

Figure 1.
The distribution of high-risk tissue at a mid-sagittal section of a vertebra with bone volume fraction (BV/TV) of 13.9%, for disc versus PMMA loading (compressive tissue-level failure is blue, and tensile tissue-level failure is red). Scale bar: 5 mm [29].

was replaced with PMMA, the anterior shift of high-risk tissue was much more appreciable. These results suggest that forward flexion loading does not appreciably change the spatial distribution of stress within the vertebral body when a compliant disc is presented adjacent to the vertebral body. The occurrence of anterior wedge fractures in cadaveric experiments in which forward bending forces were applied via a stiff material (e.g., PMMA) does not replicate the in vivo situation in human spine where compliant physiological discs exist.

The results from those high-resolution microCT-based finite element models indicate that the stress distribution of vertebral bone tissue, or at least the distribution of the most highly stressed bone tissues, is insensitive to applied compression versus forward flexion loading. This insensitivity could be explained by a simple beam-on-elastic-foundation model, in which the endplate and disc together behave as a "flexible" beam resting on an elastic foundation of trabecular bone [29]. Therefore, maximum stresses in the trabecular foundation occur beneath the center of the applied load. For uniform compression, the center of the applied load is in the central region of the vertebra, while for forward flexion, it is just slightly anterior to the center of the vertebral body.

The above mechanism applies to the loading of the endplates via a compliant disc but does not apply for the loading via the stiff PMMA. The PMMA and endplate together behaves like a "rigid" beam on the elastic foundation of trabecular bone. Thus, maximum stresses in the vertebral bone occur in the most anterior bone for forward flexion loading (**Figure 1**). Since PMMA may never represent any real disc in human spine anterior wedge fractures might not directly happen given a forward flexion loading. However, for the patients who undergo artificial disc replacement or spine fusion surgery, where their disc space is filled with a stiff material like PMMA, the adjacent vertebral bodies may be at an increased risk of wedge fracture when loaded even with a moderate degree of forward flexion.

2.3 Effects of the intervertebral disc on vertebral bone stress

Anterior wedge fractures are thought to be associated with forward flexion loading, under which the bone tissues in the anterior potion of the vertebral body are at the highest risk of initial failure. However, previous cadaver experiments have shown that bone failure occurs typically in the central regions of the trabecular bone and endplates of the vertebral body, regardless of uniform compression or forward flexion [30, 31]. As detailed above, high-resolution finite element modeling of the vertebral motion segment with a compliant intervertebral disc (elastic modulus = 8 MPa) has shown that the high-risk bone tissues in the vertebral bone are distributed primarily in the central regions of the trabecular bone and endplates for both compression and forward flexion loading [29]. Only when the flexion loading is applied to the vertebra through a stiff layer of PMMA (elastic modulus = 2500 MPa), most highly stressed bone tissues are located in the anterior aspect of the vertebra. These results imply that the material properties of the intervertebral disc may influence the distribution of vertebral stress. The questions are then: can typical variations in disc properties lead to an anterior wedge fracture? What is a typical range of disc material property?

It is known that alterations in the material properties and morphometry of the disc are associated with aging and degeneration [32, 33]; disc degeneration and loss of the height lead to an elevated risk of vertebral fractures [34]. However, it is unknown whether typical variations in the overall mechanical properties of a disc can affect either the location of high-risk tissues within the vertebra or the magnitude of vertebral stress.

To answer those questions, mechanical testing has been conducted on 16 individual whole discs from cadavers (66 ± 16 year old; mean ± SD) to measure a homogenized "effective" linear elastic modulus of the entire disc [35]. The measured elastic modulus of whole discs and the disc height were then input and varied parametrically in micro-CT-based finite element models (up to 80 million elements each) of T9 human vertebrae. The vertebral models were then virtually loaded under moderate forward flexion. The changes in stress or high-risk tissue distribution were determined as a function of the effective modulus and the height of the intervertebral disc.

Across all disc specimens, the measured effective modulus of the intervertebral disc ranged from 5.8 to 42.7 MPa; the average disc height ranged from 2.9 to 9.3 mm. Based on experimental measures of whole disc modulus [35, 36], it appears that the effective modulus of human intervertebral discs does not exceed about 100 MPa.

When the disc effective modulus increased and the disc height decreased across those measured typical ranges, the vertebral bone stresses increased but their spatial distribution was largely unchanged (**Figure 2**). Most of the high-risk tissues appeared in the central trabecular bone and endplates of the vertebra. Therefore, it can be concluded that for a moderate degree of kinematically imposed forward flexion loading, typical stiffening (increasing in disc effective modulus) or narrowing (decreasing in height) of the disc can increase the overall stress level within the vertebral body but may not lead to an anterior failure of the vertebral body.

The spatial distribution of high-risk tissue within the vertebral body is insensitive to typical variations in the effective modulus or height of the adjacent intervertebral disc. This can be explained by beam-on-elastic-foundation theory. Since high anterior stress does not develop in the vertebral bone for moderate forward flexion across the range of typical disc properties, typical variations of disc properties or height may not be directly related to a wedge-shaped fracture.

The sensitivity of the stress magnitude within the vertebral body to the effective modulus and height of the disc may have clinical implications for fracture

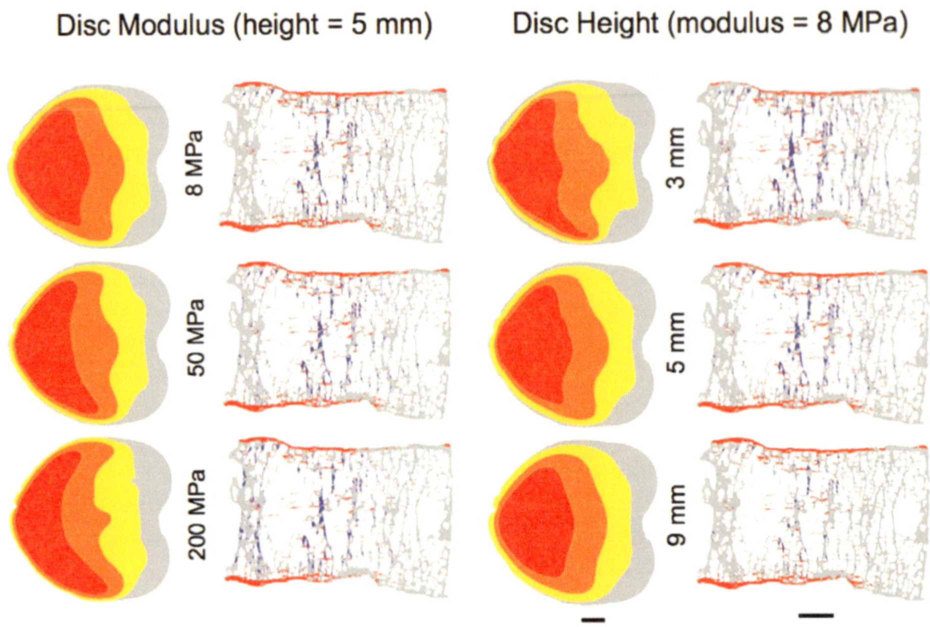

Figure 2.
Variations in the spatial distribution of axial compressive stress at a mid-transverse section of the superior disc (colors denote quartiles, red being the highest loaded, and gray the least) and the distribution of high-risk tissue at a midsagittal section of the vertebra (red and blue indicate the presence of high-risk tissue in tension and compression, respectively), for a typical range of values of the effective modulus and height of the disc. Scale bars: 5 mm [35].

risk assessment. Several studies have shown that clinical CT-based (about 1-mm resolution) finite element analysis of the human spine can predict new vertebral fractures both for women and men [37, 38] better than what DXA or quantitative CT-measured BMD can do. Those CT-based finite element analyses generally employed a compressive loading condition for all vertebrae, and loads were applied through a stiff layer of PMMA instead of a physiological disc. Since the magnitude of vertebral stress or fracture risk of vertebra is related with the material property of the disc, it can be implied that further improvements to this type of clinical finite element analysis might be needed by using actual patient-specific values of disc effective modulus. Obtaining accurate patient-specific values of disc modulus has been difficult, but this may be an area of future research as far as fracture prediction is concerned.

2.4 Possible mechanisms for spine wedge fractures

It has been shown that forward flexion loading may not directly result in an anterior wedge fracture. Aging or degeneration-related changes in disc material property and morphometry would not alter the central distribution of high-risk tissue within the vertebral body and thus may not cause wedge fractures. Therefore, the etiology of wedge fractures may lie somewhere else.

Adams [28] proposed that, with disc narrowing, habitual erect standing can lead to anterior unloading as contact occurs largely at the facet joints, and the load is transferred more through the neural arch. Such anterior unloading may cause stress shielding and adaptive bone loss of the anterior portion of the vertebral body, thus compromising the strength of the anterior bone. When a forward flexion load acts on the vertebral body, the anterior portion would fail first and thus a wedge-shaped fracture would occur.

Alternatively, it is possible that moderate forward flexion loading is not directly related with wedge fractures instead more severe forward flexion loading is. Some experiments that used a greater degree of forward flexion have found anterior wedge fractures, regardless of the state of disc degeneration [39–41].

Another possible explanation is that the modest forward bending produces initial vertebral fractures primarily in the endplates and their underlying central trabecular bone [24]. Subsequent cyclic bending loading and perhaps creep can cause progressive collapse into the anterior vertebral body. In that case, the observed morphology of the wedge-shaped fracture may reflect only the end result of the entire fracture process. It is possible that the disc may behave stiffer when its height decreases with degeneration or when the disc is loaded at a high rate, leading to increased stress within the central vertebral body, which could eventually propagate into an anterior wedge-shaped fracture. This observation could explain why degenerated-related disc space narrowing is often related with an increased risk of vertebral fractures regardless of fracture types [34].

In addition, vertebral fractures may be related to fatigue damage of bone tissues under cyclic loading [11]. Fracture might be the end of a gradual process of cumulative "fatigue failure" of the vertebral body. Or, fractures may occur slowly under constant load by gradual "creep" deformation. Clearly, further research is required to have a deeper understanding of mechanisms of spine wedge fracture.

2.5 Effects of bone tissue mechanical behavior on whole vertebral strength

The post-yield ductility of bone tissue is a type of tissue-level mechanical behavior. Bone tissue ductility is associated mainly with organic components and enables the bone tissue to deform and take load beyond the elastic range. One poorly understood multi-scale biomechanical issue is how tissue-level post-yield ductility affects the organ-level strength of the vertebral body. This multi-scale relation is also of interest clinically as tissue-level ductility can be very low in some bone pathologies, such as osteogenesis imperfecta.

This multi-scale biomechanics problem is very challenging for structurally complex vertebrae that contain both trabecular bone and cortices. It is difficult to relate any changes in tissue-level post-yield ductility to mechanical behavior (e.g., strength) of a whole vertebra. Nonlinear finite element analyses based on high resolution microCT images are well suitable to address these challenges since those large-scale finite element models can contain fine details of bone microarchitecture as well as tissue-level mechanical behaviors. Studies have been performed to investigate how whole vertebral strength is changed when the tissue-level post-yield deformation is varied from being fully ductile to fully brittle [42]. Computational simulations make it feasible to quantify the effects of tissue-level ductility on whole vertebral strength in a repeated measures manner, which is not possible only with experimentation.

For each finite element model of vertebra, two separate nonlinear finite element analyses can be performed to simulate the fully brittle and fully ductile tissue-level failure behaviors. For the fully ductile behavior, tissue-level failure is assumed by yielding; the bone tissue can only yield, it never fractures, and there is no limit on the magnitude of the post-yield tissue-level strains. For the fully brittle behavior, tissue-level fracture is assumed to occur once the yield stress (in either tension or compression) is exceeded. The elements in the computational models will be removed once they are fractured (or their yield stresses are exceeded). For all other factors, including tissue-level elastic modulus and yield stress, held fixed. For each vertebra, a finite element model can be generated (60–82 μm element size; up to 120 million elements) and virtually loaded in uniform compression. Results have shown that changing the bone tissue behavior from fully ductile to fully brittle reduced whole vertebral strength by about 40%. At overall structural failure, there was 5–10 times less failed tissue for the fully brittle than fully ductile cases. That being said, the whole vertebra is substantially strengthened when the underlying tissue is more ductile as increased ductility enables initially yielded bone tissue to continue to support the external loads. Conversely, when the bone tissue is more brittle, many of the trabeculae stop contributing to overall load bearing as the overall structure failure quickly occurs after fracture of the initially failed tissue and the lack of alternative viable load paths. These multi-scale biomechanics studies indicate that the strength of the vertebral body is determined by both bone mass and tissue-level ductility or the extent to which the bone tissue can deform beyond the tissue-level elastic range without fracturing or developing appreciable cracking or damage.

3. Biomechanical CT-based spine fracture risk prediction

3.1 QCT-based finite element analysis

DXA scanning is the clinical standard for vertebral strength (or related fracture risk) assessment. However, DXA as an imaging modality is limited due to its two-dimensional nature and its inability to differentiate material and geometric features. Quantitative computed tomography (QCT), being three-dimensional, overcomes these limitations. However, being an imaging modality that only describes bone density and geometry, it cannot describe biomechanical properties of the vertebra, an attribute that is obviously desirable for bone strength assessment. QCT-based "voxel" finite element models of the vertebral body [43] can be generated directly from QCT scans by converting voxels of the images to hexahedron elements and can be subjected to any loading conditions. Effectively, these models integrate all the information in QCT scans in a biomechanically meaningful manner and therefore promise to overcome all limitations associated with both DXA and QCT. This is so-called "biomechanical CT", a concept first proposed by Keaveny [44].

An endpoint clinical tool may use a strategy in which the QCT scan is converted, voxel by voxel, directly into a finite element model. This "voxel-based" finite element modeling technique uses the QCT voxel grayscale values and dimension data to automatically develop a finite element mesh for the region of interest [45]. Alternatively, smooth-meshed models with tetrahedral elements can be created [46, 47]. Finite element in the model is assigned local material properties based on the calibrated gray-scale information in the CT scan. Such material property-density relations have been shown to follow pow laws and are typically derived from cadaver experiments [48]. The finite elements themselves can be hexahedron or voxels, can be tetrahedral or curved, and can employ either linear or quadratic nodal-displacement formulations; special treatment of the thin cortical shell via the use of shell elements may also be implemented. Different loading conditions typical of habitual activities or more spurious overloads can be applied depending on the clinical application. Outcomes of QCT-based finite element analysis include vertebral strength, load-strength ratio, and fracture patterns and locations. These outcomes focus on the overall structure and biomechanics of bone rather than simply bone mineral density—which is most appropriate given that a bone fracture represents a biomechanical event in which external loads applied to the bone have exceeded the strength of the bone.

Validation of the QCT-based finite element model can be done by comparing model output data with measured values in mechanical testing of cadaver vertebrae.

There are several ways to validate the accuracy of the model. For example, strains at the bone surfaces of the vertebra can be measured by strain gauges and compared with the predicted strain values of the model. Alternatively, digital image/volume correlation can be performed to measure deformations of both surface bone and internal structures of the vertebra [49]. More often, apparent-level strength and stiffness measured from cadaver tests are used for validating the finite element models.

Cadaver studies have shown that QCT-based finite element analysis provides a better estimation of vertebral strength than BMD, as measured by either DXA or QCT alone [50–55]. R^2 values for prediction of experimentally measured whole vertebral compressive strength based on QCT-finite element analysis are in a range from ~0.8 to 0.9, and the slope between mechanical test-measured and finite element-predicted strength values can be very close to 1 [50–53]. However, it should be noted that QCT-based finite element analysis works well in predicting vertebral stiffness or strength for the uniform compression loading condition but has not performed as well for anterior bending [54, 55]. For example, one study found that vertebral strength in anterior bending was moderately predicted by QCT-based finite element analysis (R^2 = 0.34–0.40), which, however, was still better than QCT-based BMD (R^2 = 0.14–0.22) [54]. Another study also found moderate correlations between measured stiffness or strength and BMD (R^2 = 0.27 or 0.34) when the vertebrae were tested under anterior bending. Although QCT-based finite element analysis improved those stiffness or strength predictions appreciably (R^2 = 0.49 or 0.79), the correlations for anterior bending were not as good as those for uniform compression [55]. This may suggest different failure mechanisms of the vertebral body when loaded in uniform compression versus anterior bending. Recent studies have used microCT-based digital volume correlation to validate the accuracy of QCT-based finite element analysis in predicting vertebral failure patterns for compression and anterior flexion [49].

Currently, QCT-based finite element modeling of human vertebrae has been improved by addressing different types of loading, for example, compression and forward flexion, and the effect of intervertebral discs on vertebral strength assessment [56–58]. However, compression versus forward flexion loading has been shown to have a minor effect on vertebral stress distribution based on the observations from high-resolution microCT-based finite element analysis [29]. Variations in disc properties only affect the magnitude of the overall stress within the vertebral body, and thus, disc may serve as a critical variable that needs to be considered in the QCT-based finite element models in the future [35]. Continuing research in this field would advance the clinical use of QCT-based finite element analysis, particularly in predicting risk of osteoporotic fractures.

3.2 Vertebral fracture risk prediction based on QCT-Finite Element Analysis

Based on validation and verification with cadaver studies, QCT-based finite element analysis has been applied in clinic to predict vertebral fracture risk and evaluate osteoporosis treatment. This type of biomechanical model has been shown to improve fracture risk prediction clinically compared to what is currently possible using DXA and QCT.

The QCT-based finite element analysis was first applied to assess vertebral bone strength in live patients in a clinical research study in the early 1990s and used since then in many orthopedic biomechanics laboratory research studies [59]. There is a modest body of earlier work on finite element modeling of the vertebra that already showed great potential. In a seminal study, Faulkner et al. [59] tested the hypothesis that patient-specific bone distribution information contained in QCT voxel-based finite element models of lumbar vertebral bodies could more accurately estimate vertebral strength than bone density alone. They built voxel-based finite element models of the vertebrae taken from QCT scans of actual patients and confirmed their hypothesis by demonstrating that the models were able to better discriminate (retrospectively) osteoporotic versus nonosteoporotic patients. Further, they presented evidence that the finite element modeling technique has more discriminatory power at fracture risk assessment than measures of BMD. Bozic et al. [60] studied the effects of axial compressive loading on cervical vertebrae using a voxel-based finite element model. The model mechanistically confirmed the fracture initiation site and patterns found in clinical burst fractures of cervical vertebrae. Homminga et al. [26] used voxel-based models to study differences in the load distributions between healthy versus osteoporotic vertebrae. They showed that there was about 16% of the trabecular bone at risk of fracture in osteoporotic vertebrae versus about 1% in healthy vertebrae. Since then, many studies have demonstrated that bone strength derived from QCT-based finite element model is able to discriminate osteoporotic versus nonosteoporotic patients more effectively than BMD derived from DXA and QCT [61–63]. More importantly, QCT finite element models can predict new clinical vertebral fractures in both men and women better than DXA or QCT alone [37, 38, 64, 65]. This technique has shown a great potential in clinical application for fracture risk prediction of both spine and hip osteoporotic fractures and it is currently in clinical trials in US.

4. Conclusion

Osteoporotic spine fracture is a global issue affecting a great percentage of population, especially for the elderly. Since facture is a biomechanical event, a better understanding of the multi-scale biomechanics of the osteoporotic spine fracture would help develop better tools to improve fracture risk prediction and eventually prevent osteoporotic fractures. With the advent of micro-CT finite element modeling, studies have been performed to reveal the mechanisms of tissue deformation and microstructural failure within the entire vertebral body, information that provides insights into the micromechanics of the human vertebral body. While DXA works quite well for the hip at fracture risk prediction, it is far less successful for the spine and there is a need for improvement. The QCT-based finite element analysis promises such improvement since it is now possible to develop high-fidelity finite element models, clinically, on a patient-specific basis.

Acknowledgements

Funding from the National Natural Science Foundation of China (11702008), Beijing Excellent Talents Funds (2017000020124G277), and Support Plan for High-level Faculties in Beijing Municipal Universities (CIT&TCD201804011) is greatly acknowledged.

Conflict of interest

The author has no conflict of interest to declare.

Author details

Haisheng Yang
Department of Biomedical Engineering, College of Life Science and Bioengineering, Beijing University of Technology, Beijing, China

*Address all correspondence to: haisheng.yang@bjut.edu.cn

References

[1] Curtis EM, Moon RJ, Harvey NC, Cooper C. The impact of fragility fracture and approaches to osteoporosis risk assessment worldwide. Bone. 2017;**104**:29-38

[2] Hernlund E et al. Osteoporosis in the European Union: Medical management, epidemiology and economic burden: A report prepared in collaboration with the International Osteoporosis Foundation (IOF) and the European Federation of Pharmaceutical Industry Associations (EFPIA). Archives of Osteoporosis. 2013;**8**:136

[3] O'Neill TW, Silman AJ. Definition and diagnosis of vertebral fracture. The Journal of Rheumatology. 1997;**24**:1208-1211

[4] O'Neill TW, Felsenberg D, Varlow J, et al. The prevalence of vertebral deformity in European men and women: The European vertebral osteoporosis study. Journal of Bone and Mineral Research. 1996;**11**:1010-1018

[5] Melton LJ 3rd, Thamer M, Ray NF, et al. Fractures attributable to osteoporosis: Report from the National Osteoporosis Foundation. Journal of Bone and Mineral Research. 1997;**12**:16-23

[6] USDHHS. Aging America: Trends and Projections. 1991

[7] Akkawi I, Zmerly H. Osteoporosis: Current concepts. Joints. 2018;**6**(2):122-127

[8] Cummings SR, Bates D, Black DM. Clinical use of bone densitometry—Scientific review. Journal of the American Medical Association. 2002;**288**:1889-1897

[9] Schuit SC, van der Klift M, Weel AE, de Laet CE, Burger H, Seeman E, et al. Fracture incidence and association with bone mineral density in elderly men and women: The Rotterdam Study. Bone. 2004;**34**:195-202

[10] Christiansen BA, Bouxsein ML. Biomechanics of vertebral fractures and the vertebral fracture cascade. Current Osteoporosis Reports. 2010;**8**:198-204

[11] Adams MA, Dolan P. Biomechanics of vertebral compression fractures and clinical application. Archives of Orthopaedic and Trauma Surgery. 2011;**131**:1703-1710

[12] Berg-Johansen B, Fields AJ, Liebenberg EC, Li A, Lotz JC. Structure-function relationships at the human spinal disc-vertebra interface. Journal of Orthopaedic Research. 2018;**36**(1):192-201

[13] Edwards WT, Zheng YG, Ferrara LA, Yuan HA. Structural features and thickness of the vertebral cortex in the thoracolumbar spine. Spine. 2001;**26**:218-225

[14] Zhao FD, Pollintine P, Hole BD, Adams MA, Dolan P. Vertebral fractures usually affect the cranial endplate because it is thinner and supported by less-dense trabecular bone. Bone. 2009;**44**:372-379

[15] Silva MJ, Wang C, Keaveny TM, Hayes WC. Direct and computed-tomography thickness measurements of the human lumbar vertebral shell and end-plate. Bone. 1994;**15**:409-414

[16] Eswaran SK, Gupta A, Adams MF, Keaveny TM. Cortical and trabecular load sharing in the human vertebral body. Journal of Bone and Mineral Research. 2006;**21**:307-314

[17] Singer K, Edmondston S, Day R, Breidahl P, Price R. Prediction of thoracic and lumbar vertebral body compressive strength—Correlations

with bone-mineral density and vertebral region. Bone. 1995;**17**:167-174

[18] Fields AJ, Eswaran SK, Jekir MG, Keaveny TM. Role of trabecular microarchitecture in whole-verterbal body biomechanical behavior. Journal of Bone and Mineral Research. 2009;**29**:1523-1530

[19] Michelson JD, Myers A, Jinnah R, Cox Q, Van Natta M. Epidemiology of hip fractures among the elderly. Risk factors for fracture type. Clinical Orthopaedics and Related Research. 1995;**311**:129-135

[20] Bachmann KN, Bruno AG, Bredella MA, et al. Vertebral strength and estimated fracture risk across the BMI spectrum in women. Journal of Bone and Mineral Research. 2016;**31**(2):281-288

[21] Mosekilde L, Mosekilde L. Sex differences in age-related changes in vertebral body size, density and biomechanical competence in normal individuals. Bone. 1990;**11**:67-73

[22] Adams JE. Advances in bone imaging for osteoporosis. Nature Reviews. Endocrinology. 2013;**9**(1):28-42

[23] Van Rietbergen B, Weinans H, Huiskes R, Odgaard A. A new method to determine trabecular bone elastic properties and loading using micromechanical finite element models. Journal of Biomechanics. 1995;**28**:69-81

[24] Sanyal A, Scheffelin J, Keaveny TM. The quartic piecewise-linear criterion for the multiaxial yield behavior of human trabecular bone. Journal of Biomechanical Engineering. 2015;**137**(1):0110091-01100910

[25] Liu XS, Sajda P, Saha PK, Wehrli FW, Bevill G, Keaveny TM, et al. Complete volumetric decomposition of individual trabecular plates and rods and its morphological correlations with anisotropic elastic moduli in human trabecular bone. Journal of Bone and Mineral Research. 2008;**23**:223-235

[26] Homminga J, Van-Rietbergen B, Lochmuller EM, Weinans H, Eckstein F, Huiskes R. The osteoporotic vertebral structure is well adapted to the loads of daily life, but not to infrequent "error" loads. Bone. 2004;**34**:510-516

[27] Fields AJ, Lee GL, Keaveny TM. Mechanisms of initial endplate failure in the human vertebral body. Journal of Biomechanics. 2010;**43**:3126-3131

[28] Pollintine P, Dolan P, Tobias JH, Adams MA. Intervertebral disc degeneration can lead to "stress-shielding" of the anterior vertebral body: A cause of osteoporotic vertebral fracture? Spine. 2004;**29**:774-782

[29] Yang H, Nawathe S, Fields AJ, Keaveny TM. Micromechanics of the human vertebral body for forward flexion. Journal of Biomechanics. 2012;**45**(12):2142-2148

[30] Farooq N, Park JC, Pollintine P, Annesley-Williams DJ, Dolan P. Can vertebroplasty restore normal load-bearing to fractured vertebrae? Spine. 2005;**30**:1723-1730

[31] Landham PR, Gilbert SJ, Baker-Rand HL, Pollintine P, Robson Brown KA, Adams MA, et al. Pathogenesis of vertebral anterior wedge deformity: A 2-stage process? Spine. 2015;**40**:902-908

[32] Adams MA, Pollintine P, Tobias JH, Wakley GK, Dolan P. Intervertebral disc degeneration can predispose to anterior vertebral fractures in the thoracolumbar spine. Journal of Bone and Mineral Research. 2006;**21**:1409-1416

[33] Natarajan RN, Williams JR, Andersson GB. Recent advances

in analytical modeling of lumbar disc degeneration. Spine. 2004;**29**:2733-2741

[34] Castano-Betancourt MC, Oei L, Rivadeneira F, de Schepper EI, Hofman A, Bierma-Zeinstra S, et al. Association of lumbar disc degeneration with osteoporotic fractures; the Rotterdam study and meta-analysis from systematic review. Bone. 2013;**57**:284-289

[35] Yang H, Jekir M, Davis MW, Keaveny TM. Effective modulus of the human intervertebral disc and its effect on vertebral bone stress. Journal of Biomechanics. 2016;**49**:1134-1140

[36] O'Connell GD, Johannessen W, Vresilovic EJ, Elliott DM. Human internal disc strains in axial compression measured noninvasively using magnetic resonance imaging. Spine. 2007;**32**:2860-2868

[37] Wang X, Sanyal A, Cawthon PM, Palermo L, Jekir M, Christensen J, et al. Prediction of new clinical vertebral fractures in elderly men using finite element analysis of CT scans. Journal of Bone and Mineral Research. 2012;**27**:808-816

[38] Kopperdahl DL, Aspelund T, Hoffmann PF, Sigurdsson S, Siggeirsdottir K, Harris TB, et al. Assessment of incident spine and hip fractures in women and men using finite element analysis of CT scans. Journal of Bone and Mineral Research. 2014;**29**:570-580

[39] Adams MA, Hutton WC. Prolapsed intervertebral disc. A hyperflexion injury 1981 Volvo award in basic science. Spine. 1982;**7**:184-191

[40] Granhed H, Jonson R, Hansson T. Mineral-content and strength of lumbar vertebrae-a cadaver study. Acta Orthopaedica Scandinavica. 1989;**60**:105-109

[41] Luo J, Skrzypiec DM, Pollintine P, Adams MA, Annesley-Williams DJ, Dolan P. Mechanical efficacy of vertebroplasty: Influence of cement type, BMD, fracture severity, and disc degeneration. Bone. 2007;**40**:1110-1119

[42] Nawathe S, Yang H, Fields AJ, Bouxsein ML, Keaveny TM. Theoretical effects of fully ductile versus fully brittle behaviors of bone tissue on the strength of the human proximal femur and vertebral body. Journal of Biomechanics. 2015;**48**:1264-1269

[43] Crawford RP, Cann CE, Keaveny TM. Finite element models predict in vitro vertebral body compressive strength better than quantitative computed tomography. Bone. 2003;**33**:744-750

[44] Keaveny TM. Biomechanical computed tomography-noninvasive bone strength analysis using clinical computed tomography scans. Annals of the New York Academy of Sciences. 2010;**1192**:57-65

[45] Keyak JH, Meagher JM, Skinner HB, Mote CD Jr. Automated three-dimensional finite element modelling of bone: A new method. Journal of Biomedical Engineering. 1990;**12**:389-387

[46] Yang H, Ma X, Guo T. Some factors that affect the comparison between isotropic and orthotropic inhomogeneous finite element material models of femur. Medical Engineering & Physics. 2010;**32**:553-560

[47] Zysset P, Pahr D, Engelke K. Comparison of proximal femur and vertebral body strength improvements in the FREEDOM trial using an alternative finite element methodology. Bone. 2015;**81**:122-130

[48] Zioupos P, Cook RB, Hutchinson JR. Some basic relationships between density values in cancellous and

cortical bone. Journal of Biomechanics. 2008;**41**:1961-1968

[49] Jackman TM, DelMonaco AM, Morgan EF. Accuracy of finite element analyses of CT scans in predictions of vertebral failure patterns under axial compression and anterior flexion. Journal of Biomechanics. 2016;**49**:267-275

[50] Kinzl M, Schwiedrzik J, Zysset PK, Pahr DH. An experimentally validated finite element method for augmented vertebral bodies. Clinical Biomechanics (Bristol, Avon). 2013;**28**:15-22

[51] Imai K, Ohnishi I, Bessho M, Nakamura K. Nonlinear finite element model predicts vertebral bone strength and fracture site. Spine. 2006;**31**:1789-1794

[52] Buckley JM, Loo K, Motherway J. Comparison of quantitative computed tomography-based measures in predicting vertebral compressive strength. Bone. 2007;**40**:767-774

[53] Dall'Ara E, Pahr D, Varga P, Kainberger F, Zysset P. QCT-based finite element models predict human vertebral strength in vitro significantly better than simulated DEXA. Osteoporosis International. 2012;**23**:563-572

[54] Buckley JM, Cheng L, Loo K, Slyfield C, Xu Z. Quantitative computed tomography-based predictions of vertebral strength in anterior bending. Spine. 2007;**32**:1019-1027

[55] Dall'Ara E, Schmidt R, Pahr D, Varga P, Chevalier Y, Patsch J, et al. A nonlinear finite element model validation study based on a novel experimental technique for inducing anterior wedge-shape fractures in human vertebral bodies in vitro. Journal of Biomechanics. 2010;**43**:2374-2380

[56] Costa MC, Tozzi G, Cristofolini L, Danesi V, Viceconti M, Dall'Ara E. Micro finite element models of the vertebral body: Validation of local displacement predictions. PLoS One. 2017;**12**(7):e0180151

[57] Hussein AI, Louzeiro DT, Unnikrishnan GU, Morgan EF. Differences in trabecular microarchitecture and simplified boundary conditions limit the accuracy of quantitative computed tomography-based finite element models of vertebral failure. Journal of Biomechanical Engineering. 2018;**140**(2):021004

[58] Maquer G, Schwiedrzik J, Huber G, Morlock MM, Zysset PK. Compressive strength of elderly vertebrae is reduced by disc degeneration and additional flexion. Journal of the Mechanical Behavior of Biomedical Materials. 2015;**42**:54-66

[59] Faulkner KG, Cann CE, Hasegawa BH. Effect of bone distribution on vertebral strength: Assessment with a patient-specific nonlinear finite element analysis. Radiology. 1991;**179**:669-674

[60] Bozic KJ, Keyak JH, Skinner HB, et al. Three-dimensional finite element modeling of a cervical vertebra: An investigation of burst fracture mechanism. Journal of Spinal Disorders. 1994;**7**:102-110

[61] Imai K, Ohnishi I, Yamamoto S, Nakamura K. In vivo assessment of lumbar vertebral strength in elderly women using computed tomography-based nonlinear finite element model. Spine. 2008;**33**:27-32

[62] Melton LJ, Riggs BL, Keaveny TM, et al. Structural determinants of vertebral fracture risk. Journal of Bone and Mineral Research. 2007;**22**:1885-1892

[63] Graeff C, Marin F, Petto H, et al. High resolution quantitative computed tomography-based assessment of trabecular microstructure and strength

estimates by finite-element analysis of the spine, but not DXA, reflects vertebral fracture status in men with glucocorticoid-induced osteoporosis. Bone. 2012;**52**(2):568-577

[64] Zysset P, Qin L, Lang T, et al. Clinical use of quantitative computed tomographybased finite element analysis of the hip and spine in the management of osteoporosis in adults: The 2015 ISCD Official Positions-Part II. Journal of Clinical Densitometry. 2015;**18**(3):359-392

[65] Johannesdottir F, Allaire B, Bouxsein ML. Fracture prediction by computed tomography and finite element analysis: Current and future perspectives. Current Osteoporosis Reports. 2018;**16**(4):411-422

Chapter 6

Trapeziometacarpal Joint: A Mechanical Explanation of Total Prosthesis Failures

Victoria Spartacus

Abstract

Total prosthesis, which preserves strength and respects trapeziometacarpal (TMC) joint range of motion, is a surgical option considered for the TMC joint replacement. With the usual ball-and-socket design, patients obtain faster pain relief, stronger grip function, and shorter convalescence than with trapeziectomy. However, prostheses currently used have led to various early complications, especially in active and young patients. Revisions are most often due to the loosening of the trapezium cup and of the metacarpal stem and to the luxation of the first metacarpal bone. The short lifespan of these devices suggests the difficulty of designing a prosthesis that respects the complex anatomy and motions of the TMC joint. Early implant failure may reflect the fact that the current devices do not exactly replicate the true kinematics. The aim of this chapter will be to shed light mechanical explanations for TMC prosthesis failures in regard to the complex kinematics of the joint.

Keywords: trapeziometacarpal joint, total prosthesis, failures

1. Introduction

The thumb column is composed of three bones: the scaphoid, trapezium, and first metacarpal (M1) (**Figure 1**).

The trapeziometacarpal (TMC) joint, composed of the trapezium and the M1, plays a capital role in strength and movement orientation of this column. This joint allows the opposition of the thumb and thus is highly solicited.

Arthrosis of the TMC joint, called rhizarthrosis, is the second most frequent arthritis of the hand [1]. It is prevalent among middle-aged and postmenopausal women [2]; one third of women over 55 are radiographically affected by rhizarthrosis [3]. This painful and disabling pathology limits the range of motion and the strength of the thumb. It not only occurs mainly on the cartilage of the trapezium and the first metacarpal joint but can also affects the whole cartilaginous surface of the trapezium as defined by Dell [4] and Eaton and Littler [5] (**Figure 2**).

Its natural progression typically results in less pain, but hand function is reduced because of the Z-deformity of the thumb column. The base of the first metacarpal (M1) dislocates into adduction, in combination with hyperextension of the metacarpophalangeal (MCP) joint and contracture of the first web space. This natural progression toward pain relief often results in patients and surgeons taking in a wait-and-see approach. When conservative treatments fail, surgical options can be considered. The numerous surgical solutions proposed include trapeziectomy with ligament interposition and arthrodesis [6]. Arthrodesis involves fusion of the trapezium and M1, to reduce pain by restraining movement. In trapeziectomy, the trapezium bone is removed and all the ligaments are sectioned. Trapeziectomy yields good pain results, but patients lack strength during movements [7]. These treatments do not respect the original anatomy of the joint and modify the kinematics of the TMC joint.

Another and more recent surgical option is total prosthesis, which preserves strength and respects TMC joint kinematics. With the usual ball-and-socket design, patients obtain faster and better pain relief, stronger grip function, and shorter convalescence than with trapeziectomy [8, 9]. However, there are also many reports of poor results [10]. Revisions are most often due to the loosening of the trapezium cup [11, 12] and of the metacarpal stem and to the luxation of the first metacarpal bone [13] (**Figure 3**).

The prostheses currently used have thus led to various early complications, especially in active young patients. The short lifespan of these devices suggests the difficulty of designing a prosthesis that respects the complex anatomy and motions of the TMC joint. Early implant failure may reflect the fact that current devices do not exactly replicate the true kinematics. Improved knowledge of TMC kinematics with implant could also enhance the design and consequently the lifespan of implants.

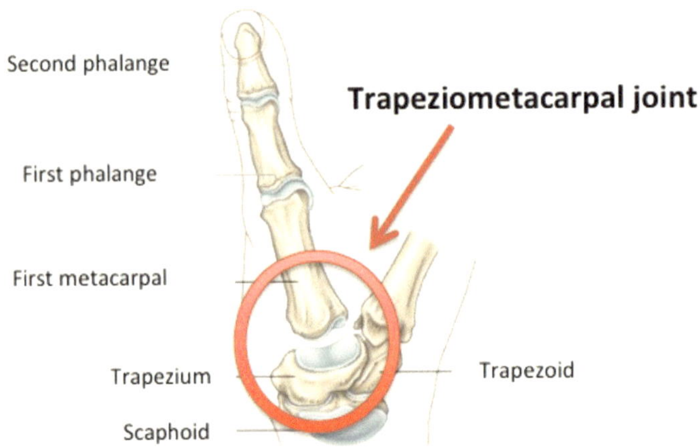

Figure 1.
Thumb column.

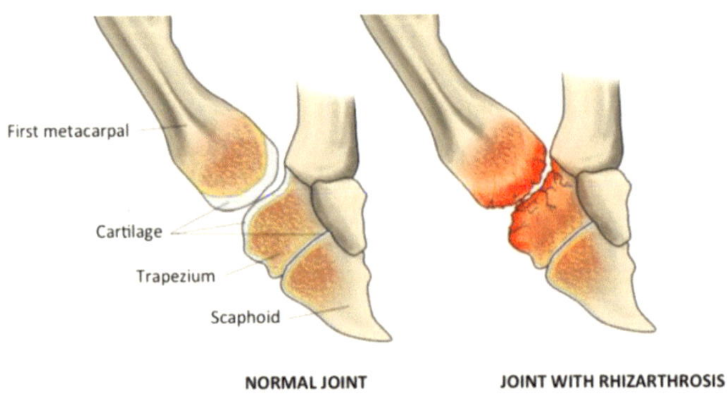

Figure 2.
Rhizarthrosis of the trapeziometacarpal joint.

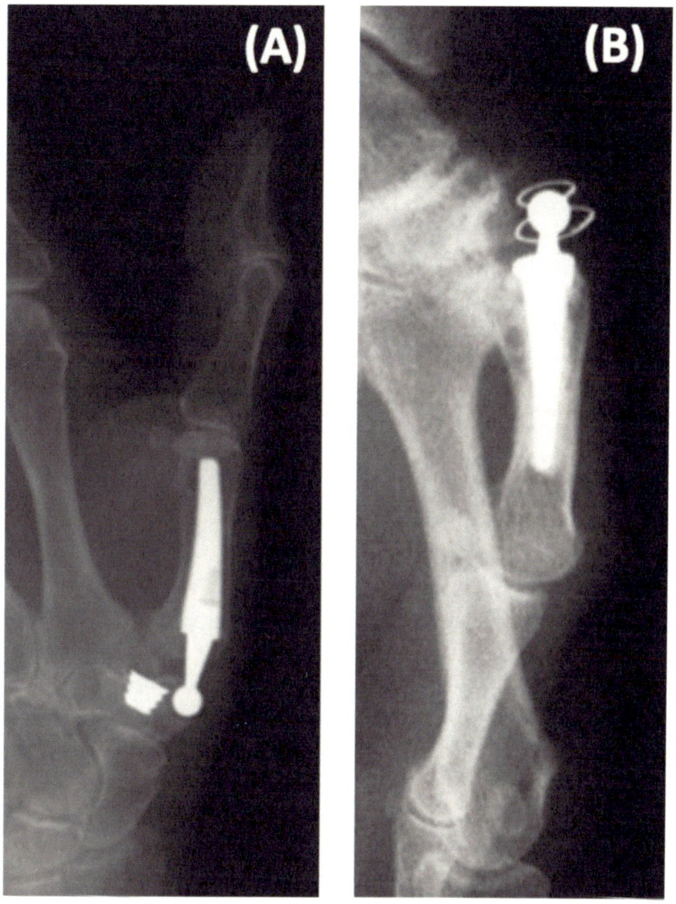

Figure 3.
(A) Luxation of the first metacarpal bone by Klahn et al. [14] and (B) loosening of the trapezium cup by Lemoine et al. [15].

2. Trapeziometacarpal joint kinematics

A precise description of the TMC joint in an anatomic plane is complicated by the complexity of the joint anatomy. The TMC joint is 40° orientated with respect to the three anatomic planes (**Figure 4**). This joint is described according to four principal views:

- Palmar view: anterior view, observation of the palm of the hand

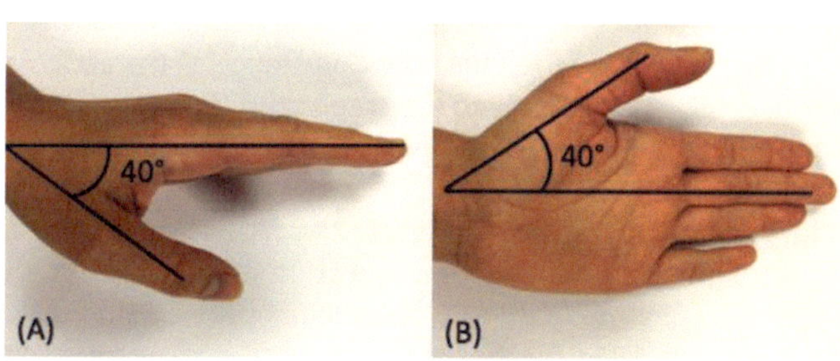

Figure 4.
Thumb in resting position of a left hand. (A) Frontal plane and (B) sagittal plane.

- Dorsal view: posterior view, observation in the dorsal face of the hand
- Ulnar view: a medial viewing angle (ulna side)
- Radial view: a lateral viewing angle (radius side)

In order to well understand the hand anatomy descriptive documents, certain criteria have to be taken into account. From the general view, the body is divided into two parts by vertical column. Then in anatomic study, the more the part of the body is close to the column, the more this part is medial. The more distant it is becoming from the column, the more lateral it is becoming from it. However, a difficulty arises with the human hand. The prono-supination movement gives to the thumb both possibilities according to the articular position of the wrist and the elbow. Thus a reference position has been established also called resting position. This position corresponds to the forearm in supination with the thumb in the lateral side, outwardly directed.

As it was mentioned previously, the TMC joint plays a fundamental role in the opposition mechanism of the thumb. It imparts a very high mobility, in regard to the M1 movement related to the trapezium. The TMC joint increasingly used, particularly with the appearance of the new technology, was a subject of interest.

The TMC joint is considered as a saddle-shape joint with two degrees of freedom (DOF) with nonorthogonal and nonintersected axes. This joint allows the thumb's different movements: flexion/extension, abduction/adduction, axial rotation, and circumduction (**Figure 5**):

- Flexion: thumb movement in medial direction. The thumb fingertip draws near the palm of the hand.
- Extension: thumb movement in posterolateral direction. Opposite side of flexion.
- Abduction: thumb movement in anterolateral direction, perpendicular to the palm of the hand.
- Adduction: thumb movement in posteromedial direction, in the palm of the hand plane.
- Circumduction: combination of the flexion/extension and the abduction/adduction movements.
- Axial rotation: rotation of the first metacarpal around his longitudinal axis.

The TMC joint is also a noncongruent joint with articular surface of the trapezium bigger than the M1 (**Figure 6**). This noncongruence of the articular surfaces implies translation of the M1 relatively to the trapezium.

Chèze et al. [16, 17] reported in their study values of M1 displacements in regard to the trapezium (**Table 1**).

Translations are considered to be negligible. Nevertheless, in regard to articular surface measurements, this conclusion is questionable [18, 19]. A study [18] realized measures on trapezium diameter. The results showed that the average diameter of the trapezium is 11.96 ± 1.32 mm. From these measurements it can be deduced that the M1 diameter is 34% larger than the trapezium diameter. In more recent study [19], the results show that the mean length of the trapezium

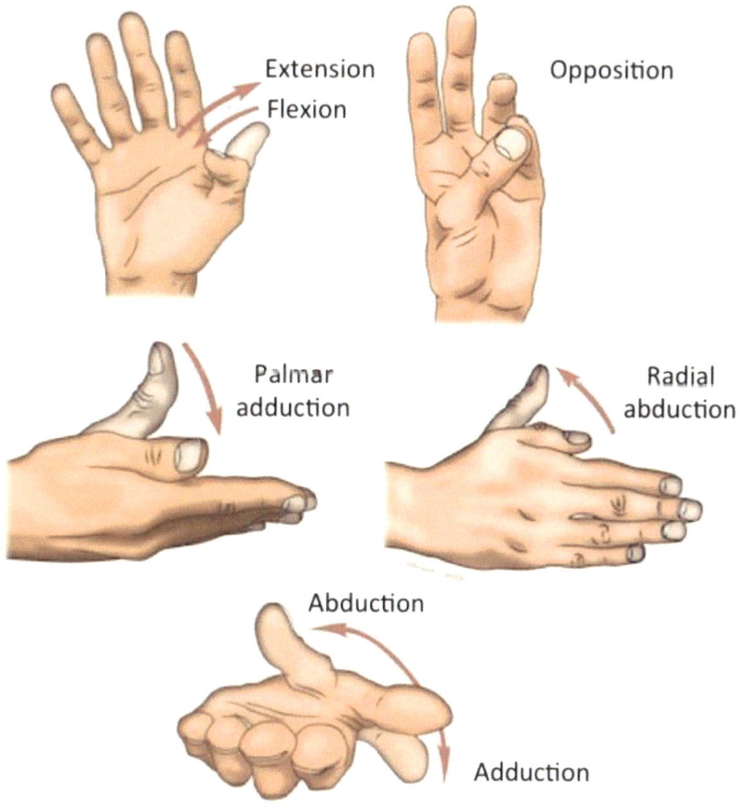

Figure 5.
Thumb movements by Dr Grégoire Chick.

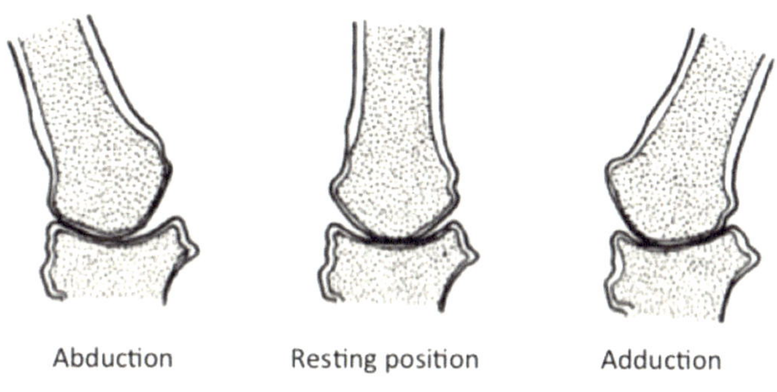

Figure 6.
Noncongruence of the TMC joint by Napier et al. [14].

is 19.2 mm. From this value the M1 lateral displacement represents 25.5% of the trapezium surface that could not be negligible. In addition, in more recent study [20], displacement of the M1 was measured for different functional tasks as key pinch, jar grasp, and jar twist. In this study, the maximum radial/ulnar displacement of the M1 is 5.7 mm. With this value the M1 displacement represents 47.7% of the trapezium diameter.

Moreover, the two DOF models are not precise enough. The dispersion in range of motion in the literature shows the difficulties to analyze with accuracy the kinematic of the TMC joint (**Table 2**).

Medial/lateral displacement	1.7 mm, maximum 4.9 mm
Proximal/distal displacement	2.0 mm, maximum 5.9 mm
Anterior/posterior displacement	3.0 mm, maximum 6.3 mm

Table 1.
Average displacement of the M1 by Chèze et al. [15]

	Kaplan [19]	Cooney et al. [20]	Chèze et al. (2001)[17]	Coert et al. [21]	Li et al. (2007) [24]	Cheema et al. (2006) [25]
Axial rotation	17.5°	16.5°	38°	90°	45°	89°

Average range of movement of axial rotation.

Table 2.
Example of disparity in range of movement of the thumb.

Several methods of thumb movement quantification have been established. These methods, including video technics [22, 23, 26], reflective markers [27, 28], and electromagnetic design [17, 29, 30], used different external markers as reference point. In 2001, Chèze et al. [17] elaborated an experimental protocol with a motion analysis system that allows to obtain the range of motion. In 2009, Goubier et al. [29] studied the movement with an optoelectronic system. These protocols used external markers to analyze the thumb kinematics. Then interference of soft tissue could influence the movements.

Tomographic imaging is an emerging method in thumb kinematics analysis. In vivo measurement has been established to evaluate axial rotation [25]. This rotation has been measured drawing reference lines in CT scan images. This study allows to measure the axial orientation of the M1.

In vivo studies, using CT scan methodology, based on TMC joint movement have been performed [31–34]. They described nonorthogonal and nonperpendicular axes, but they did not particularly described translations and axial rotation.

All previous studies aimed to characterize the thumb movement with different specific methods. However, they only considered the movement of healthy joint and not on pathologic one. Furthermore these studies did not focus on movement of the TMC prosthesis.

2.1 Material and methods

In order to analyze the movement of the TMC prosthesis, CT scans of the TMC joint under various postures of the thumb were performed with a general electric scanner light speed VCT64. The scan acquisition parameters were mAs 90, kV 120, slice thickness 0.625, and FOV small. Eight hands were obtained from six embalmed Caucasian cadaveric subjects, two males (three hands) and four females (five hands) with different degrees of rhizarthrosis according to the Dell classification [4] (**Table 3**). The subjects were divided into three groups: group 1, subjects with either no or stage 1 arthrosis; group 2, subjects with stage 2 and stage 3 arthrosis; and group 3, subjects with stage 4 arthrosis.

Three postures were chosen to cover the full range of thumb motion: commissural closing (**Figure 7(A)**), grip (**Figure 7(B)**), and opposition (**Figure 7(C)**). Using Mimics® (Materialize 3D, Belgium), the DICOM data from the CT scans were used to develop 3D reconstructions of the TMC joint.

Subject	1.R	2.R	3.R	4.R	5.R	5.L	6.R	6.L
Gender	M	M	F	F	F	F	F	F
Age	82	90	92	84	94	94	94	94
Side	Right	Right	Right	Right	Right	Left	Right	Left
Stage of arthrosis	3	1	3	4	NA	4	2	2

NA, no arthrosis; Stage 1, subchondral sclerosis with no osteophytes or subluxation; Stage 2, small subluxation and small internal osteophytes; Stage 4, no remaining space between the bones.

Table 3.
Summary of cadaveric subjects.

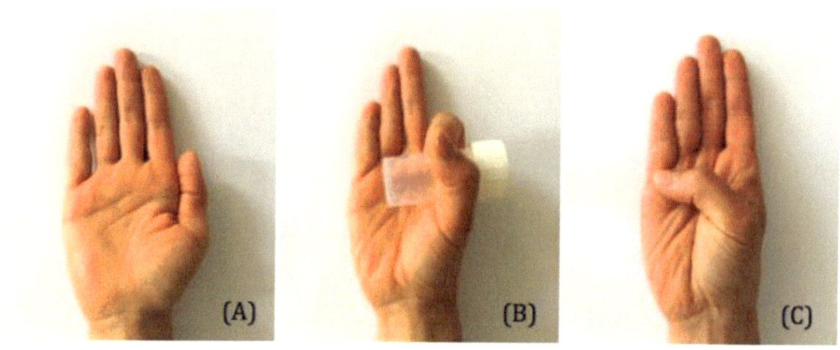

Figure 7.
(A) Thumb in commissural closing posture, (B) thumb in grip posture and (C) thumb in opposition posture.

CAD models of a currently used prosthesis (**Figure 8**) were coupled with the 3D reconstructions of the joint to provide numerical models of the TMC joint with a ball-and-socket implant. The CAD model, composed of three elements (stem, neck, and trapezium cup), was placed in the ATM using 3-Matic® (Materialize, Belgium).

Working from the 3D models, the commissural closing posture was the reference posture for the complex trapezium/cup and the complex M1/stem/neck. These complexes were superposed to the M1 and the trapezium in opposition and in grip posture, using a surface-based registration procedure based on the iterative closest point (ICP) [35] (**Figure 9**).

Thus for each posture, we determined the position of the complex M1/stem/neck relative to the complex trapezium/cup. Then, for each hand, considering the complex trapezium/cup as fixed, the different postures were superposed using the same surface-based registration procedure. The method of superposition was previously described by Cerveri et al. [31] (**Figure 10**).

Potential translations of the head relative to the cup were determined by the distance between the center of the cup and the center of the head. This distance was calculated for grip and opposition posture (**Figure 11**). Zero distance is being taken as indicating none translation. A distance other than zero was taken as a sign of translation, with the head of the neck penetrating the cup. In this case, the intersection volume between the cup and the head was calculated in order to evaluate the percentage of the cup volume occupied by the head.

2.2 Results

Regarding the straight neck, the smallest distances between the cup and the head of the prosthesis in grip posture or in opposition posture are 1.3 and 0.8 mm, respectively. Regarding the angled neck, the smallest distances between the cup

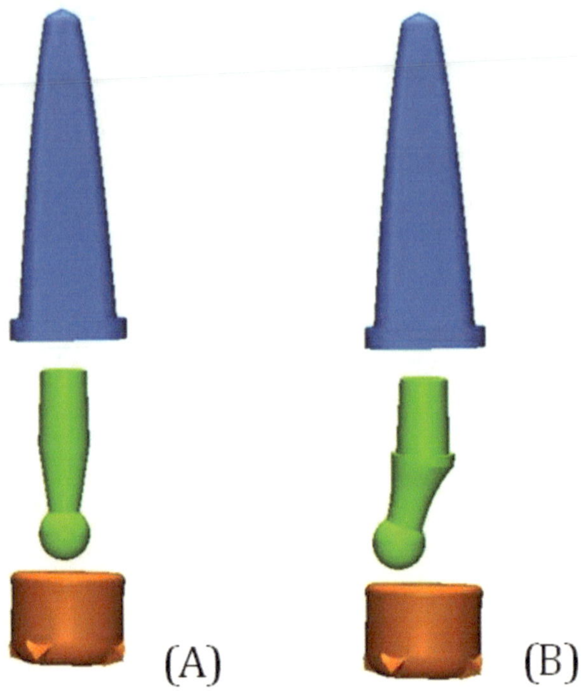

Figure 8.
CAD models of current prosthesis used in this study: cup in orange, neck in green, and stem in blue. (A) Prosthesis with a straight neck and (B) prosthesis with an angle neck.

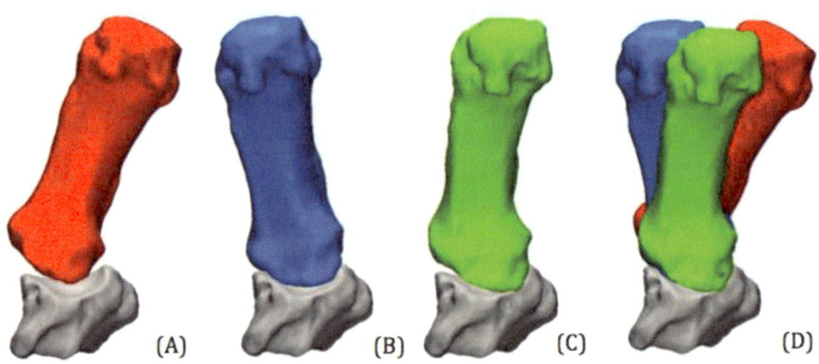

Figure 9.
Superposition procedure.

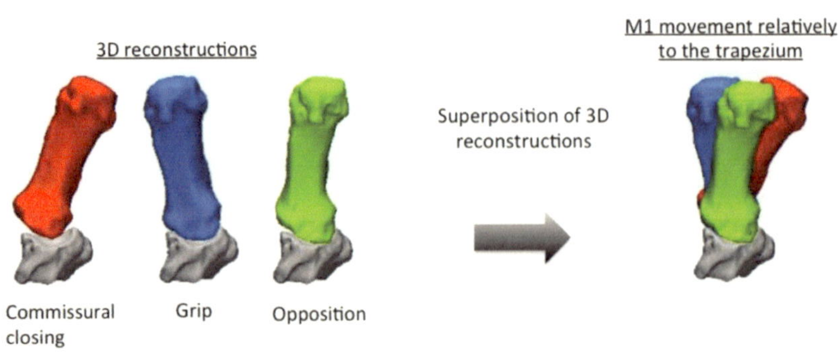

Figure 10.
Superposition procedure for complex trapezium/cup and complex M1/stem/neck.

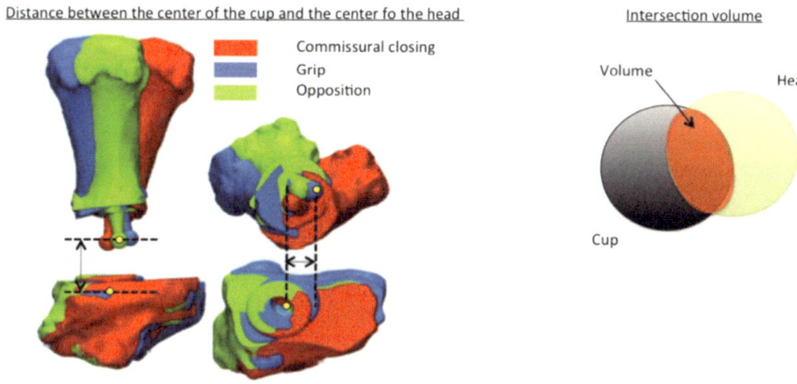

Figure 11.
Right: distance between cup and head center. Left: intersection volume between the cup and the head.

	D1 (mm)	V1 (mm^3)	D2 (mm)	V2 (mm^3)	D3 (mm)	V3 (mm^3)	D4 (mm)	V4 (mm^3)
5.R G1	5.2	0.0	2.5	6.3	3.5	0.8	4.9	0.0
2.R G1	5.3	0.0	3.8	0.2	3.4	1.0	5.4	0.0
6.R G2	2.3	9.0	0.8	24.4	1.4	16.6	2.9	3.6
6.L G2	3.2	2.6	1.6	15.3	1.5	15.7	3.6	0.5
3.R G2	4.3	0.0	3.1	2.8	3.2	1.9	5.1	0.0
1.R G2	4.9	0.0	2.0	11.1	1.8	12.4	4.0	0.0
4.R G3	1.5	16.2	0.8	24.4	0.6	25.5	1.2	19.3
5.L G3	1.3	17.9	1.4	18.0	0.7	25.1	1.5	15.7

Table 4.
For each subject, distances between the head gravity center of the straight neck and the center of the cup in grip posture (D1) and in opposition posture (D2). V1 is the theoretical intersection volume between the cup and the head of the straight neck in grip posture and V2 in opposition posture. Same results are observed for angled neck in D3, D4, V3, and V4.

and the head of the prosthesis in grip posture or in opposition posture are 0.6 and 1.2 mm, respectively.

Thus, for both types of the neck, for each posture, distances are superior to zero. Thus for each of the three groups, the head of the prosthesis translates during movements and penetrates into the cup.

Regarding prosthesis with a straight neck, the intersection volume between cup and head varied from 0 to 24.4 mm^3 (**Table 4**). When the volume is 0, the elements do not intersect. In this case, the head is completely out of the cup. When the volume is 24.4 mm^3, the elements intersect. The intersecting volume represents 67.6% of the cup volume.

Regarding prosthesis with an angled neck, the intersection volume between cup and head varied from 0 to 25.5 mm^3 (**Table 4**). When the volume is 0, the elements do not intersect. In this case, the head is completely out of the cup. When the volume is 25.5 mm^3, the elements intersect. The intersecting volume represents 70.8% of the cup volume.

2.3 Conclusion

The current total prosthesis, with the usual ball-and-socket design, preserves strength and respects TMC range of motion. However, they have led to various early complications, especially in active young patients. This study focused on the origin of the TMC prostheses failures to elucidate their mechanical explanation.

The CAD model of a ball-and-socket design prosthesis implanted in each different postures of each subject shows that the original kinematics of the joint is disturbed by the prosthetic elements. The displacements of the head of the prosthesis between each posture are greater than those of the cup. We observe numerical penetration by the head of the prosthesis into the cup. This phenomenon could in vivo correspond to an overstress of prosthetic elements. This penetration supports the fact that the most frequent prosthesis revision is due to the loosening of the trapezium cup [11, 36]. Chakrabarti et al. [36] showed that 91% of prosthesis revisions are due to loosening of the cup.

Thus, the movement of the prosthesis does not fully respect the anatomical kinematics. The TMC joint is known to have nonintersecting and nonorthogonal rotation axes [31, 32]. Moreover, this joint allowed translations of the M1 over the trapezium [32, 37]. Then simplifying the TMC joint to a ball-and-socket articulation could in vivo lead to an overstress of the prosthesis. The overstressing produced by this design could partly explain the short lifespan of current prostheses. In the light of this study's findings, a review of the design of these prostheses appears warranted.

Author details

Victoria Spartacus
University of Valenciennes and Hainaut-Cambrésis, Valenciennes, France

*Address all correspondence to: victoria.spartacus@wanadoo.fr

References

[1] Batra S, Kanvinde R. Osteoarthritis of the thumb trapeziometacarpal joint. Current Orthopaedics. 2007;**21**:135-144

[2] Stussi JD, Dap F, Merle M. A retrospective study of 69 primary rhizarthrosis surgically treated by total trapeziectomy followed in 34 cases by interpositional tendinoplasty and in 35 cases by suspensioplasty. Chirurgie de la main. 2000;**19**:116-127. PMID: 10904830

[3] Armstrong A, Hunter J, Davis T. The prevalence of degenerative arthritis of the base of the thumb in post-menopausal women. The Journal of Hand Surgery: Journal of the British Society for Surgery of the Hand. 1994;**19**:340-341. PMID: 8077824

[4] Dell PC, Brushart TM, Smith RJ. Treatment of trapeziometacarpal arthritis: Results of resection arthroplasty. Journal of Hand Surgery. 1978;**3**:243-249. PMID: 659819

[5] Eaton RG, Littler JW. Ligament reconstruction for the painful carpometacarpal joint. Clinical Orthopaedics. 1987;**220**:14-26

[6] Gay AM, Celier A, Iniesta A, Legré R. Surgery for trapeziometacarpal osteoarthritis. Hand Surgery & Rehabilitation. 2016;**35**(4):238-249. DOI: 10.1016/j.hansur.2016.06.002

[7] Degeorge B, Dagneaux L, Andrin J, Lazerges C, Coulet B, Chammas M. Do trapeziometacarpal prosthesis provide better metacarpophalangeal stability than trapeziectomy and ligamentoplasty? Orthopaedics & Traumatology: Surgery & Research. 2018;**104**(7):1095-1100. DOI: 10.1016/j.otsr.2018.07.008

[8] Dehl M, Chelli M, Lippmann S, Benaissa S, Rotari V, Moughabghab M. Results of 115 Rubis II reverse thumb carpometacarpal joint prostheses with a mean follow-up of 10 years. Journal of Hand Surgery (European Volume). 2017;**42**(6):592-598. DOI: 10.1177/1753193416687508

[9] Semere A, Vuillerme N, Corcella D, Forli A, Moutet F. Results with the Roseland HAC trapeziometacarpal prosthesis after more than 10 years. Chirurgie de la Main. 2015;**34**:59-66. DOI: 10.1016/j.main.2015.01.004

[10] Robles-Molina MJ, Lopez-Caba F, Gomez-Sanchez RC, Cardenas-Grande E, Pajares-Lopez M, Hernandez-Cortes P. Trapeziectomy with ligament reconstruction and tendon interposition versus a trapeziometacarpal prosthesis for the treatment of thumb basal joint osteoarthritis. Orthopedics. 2017;**40**(4):e681-e686. DOI: 10.3928/01477447-20170503-03

[11] Toffoli A, Teissier J. MAÏA trapeziometacarpal joint arthroplasty: Clinical and radiological outcomes of 80 patients with more than 6 years of follow-up. Journal of Hand Surgery (American Volume). 2017;**42**(10):838.e1-838.e8. DOI: 10.1016/j.jhsa.2017.06.008

[12] Caekebeke P, Duerinckx J. Can surgical guidelines minimize complications after Maïa® trapeziometacarpal joint arthroplasty with unconstrained cups? Journal of Hand Surgery (European Volume). 2018;**43**(4):420-425. DOI: 10.1177/1753193417741237

[13] Bricout M, Rezzouk J. Complications and failures of the trapeziometacarpal Maïa® prosthesis: A series of 156 cases. Hand Surgery & Rehabilitation. 2016;**35**(3):190-198. DOI: 10.1016/j.hansur.2016.02.005

[14] Klahn A, Nygaard M, Gvozdenovic R, Boeckstyns MEH. Elektra prosthesis

for trapeziometacarpal osteoarthritis: A follow-up of 39 consecutive cases. Journal of Hand Surgery. 2012;**37**E:605-609. DOI: 10.1177/1753193412443501

[15] Lemoine S, Wavreille G, Alnot JY, Fontaine C, Chantelot C, le groupe GUEPAR. Second generation guepar total arthroplasty of the thumb basal joint: 50 months follow-up in 84 cases. Orthopaedics & traumatology: Surgery & research. 2008;**95**:63-69. DOI: 10.1016/j.otsr.2008.06.001

[16] Napier JR. The form and the function of the carpo-metacarpal joint oh the thumb. Journal of Anatomy. 1955;**89**:362-369. PMID: 13251966

[17] Chèze L, Doriot N, Eckert M, Rumelhart C, Comtet JJ. In vivo cinematic study of the trapezometacarpal joint. Chirurgie de la main. 2001;**20**:23-30. PMID: 11291316

[18] Bettinger P, Berger R. Functional ligamentous anatomy of the trapezium and trapeziometacarpal joint (gross and arthroscopic). Hand Clinics. 2001;**17**(2):151-168. PMID: 11478038

[19] Loisel F, Chapuy S, Rey PB, Obert L, Parratte B, Tatu L, Lepage D. Dimensions of the trapezium bone: A cadaver and CT study. Surgical and Radiologic Anatomy. 2015;**37**(7):787-792. DOI: 10.1007/s00276-015-1418-7

[20] Halilaj E, Rainbow MJ, Got C, Schwartz JB, Moore DC, Weiss AP, Ladd AL, Crisco JJ. In vivo kinematics of the thumb carpometacarpal joint during three isometric functional tasks. Clinical Orthopaedics and Related Research. 2014;**472**(4):1114-1122. DOI: 10.1007/s11999-013-3063-y

[21] Kaplan EB. Functional and Surgical Anatomy of the Hand. 2nd ed. Philadelphia: J.B. Lippincott; 1965

[22] Cooney WP 3rd, Lucca MJ, Chao EY, Linscheid RL. The kinesiology of the thumb trapeziometacarpal joint. Journal of Bone and Joint Surgery (American). 1981;**63**(9):1371-1381. PMID: 7320028

[23] Coert JH, Hoek van Dijke GA, Hovius SER, Snijders CJ, Meek MF. Quantifying thumb rotation during circumduction utilizing a video technique. Journal of Orthopaedic Research. 2003;**21**:1151-1155. DOI: 10.1016/S0736-0266(03)00114-1

[24] Li Z-M, Tang J. Coordination of the thumb joint during opposition. Journal of Biomechanics. 2007;**40**: 502-510. PMID: 16643926

[25] Cheema TA, Cheema NI, Tayyab R, Firoozbakhsh K. Microstructural adaptation in trapezial bone due to subluxation of the thumb. Journal of Hand Surgery. 2009;**31**:76-79. DOI: 10.1002/jor.20500

[26] Kuo L-C, Su F-C, Chiu H-Y, Yu C-Y. Feasibility of using a video-based motion analysis system for measuring thumb kinematics. Journal of Biomechanics. 2002;**35**:1499-1506. PMID: 12413969

[27] Lin H-T, Kuo L-C, Liu H-Y, Wu W-L, Su F-C. The three-dimensional analysis of three thumb joints coordination in activities of daily living. Clinical Biomechanics. 2011;**28**:371-376. DOI: 10.1016/j.clinbiomech.2010.11.009

[28] Chèze L, Dumas R, Comtet J-J, Rumelhart C, Fayet M. Determination of the number of degrees of freedom of the trapeziometacarpal joint-an in vitro study. IRBM. 2012;**33**:271-276

[29] Goubier JN, Devun L, Mitton D, Lavaste F, Papadogeorgou E. Normal range of motion of trapeziometacarpal joint. Chirurgie de la main. 2009;**28**: 297-300. PMID: 19762264

[30] Kuo L-C, Cooney WP, An K-N, Lai K-Y, Wang S-M, Su F-C. Effects of age and gender on the movement workspace

of the trapeziometacarpal joint. Journal of Engineering in Medecine. Part H. 2009;**223**:133-142. PMID: 19278191

[31] Cerveri P, De Momi E, Marchente M, Baud-Bovy G, Scifo P, Barros RML, Ferrigno G. Method for the estimation of double hinge kinematic model for the trapeziometacarpal joint using MR imaging. Computer Methods in Biomechanics and Biomedical Engineering. 2010;**13**(3):387-396. PMID: 19802754

[32] Crisco J, Halilaj E, Douglas CM, Patel T, Weiss A-P, Ladd L. In vivo kinematics of the trapeziometacarpal joint during thumb extension-flexion and abduction-adduction. Journal of Hand Surgery American. 2015;**40**(2):289-296. DOI: 10.1016/j.jhsa.2014.10.062

[33] D'Agostino P, Dourthe B, Kerkhof F, Stockmans F, Vereecke EE. In vivo kinematics of the thumb during flexion and adduction motion: Evidence for screw-home mechanism. Journal of Orthopaedic Research. 2017;**35**(7): 1556-1564. DOI: 10.1002/jor.23421

[34] Wang KK, Zang X, McCombe D, Ackland DC, Ek ET, Tham SK. Quantitative analysis of in-vivo thumb carpometacarpal joint kinematics using four-dimensional computed tomography. Journal of Hand Surgery (European Volume). 2018; 1753193418789828;**0**(0):1-10. DOI: 10.1177/1753193418789828

[35] Besl PJ, McKay ND. A method for registration of 3-D shapes. IEEE Transactions on pattern analysis and machine intelligence. 1992;**14**(2):239-256

[36] Chakrabarti AJ, Robinson AHN, Gallagher P. De La Caffinière thumb carpometacarpal replacements: 93 cases at 6 to 16 years follow-up. Journal of Hand Surgery (British). 1997;**22**: 695-698. PMID: 9457567

[37] Crisco JJ, Patel T, Halilaj E, Moore DC. The envelope of physiological motion of the first carpometacarpal joint. Journal of Biomechanical Engineering. 2015;**137**(10):101002. DOI: 10.1115/1.4031117